Springer Series in Computational Mathematics

45

For further volumes:
http://www.springer.com/series/797

Wolfgang Hackbusch

The Concept of Stability in Numerical Mathematics

 Springer

Wolfgang Hackbusch
MPI für Mathematik in den
    Naturwissenschaften
Leipzig, Germany

ISSN 0179-3632
ISBN 978-3-662-51371-2   ISBN 978-3-642-39386-0 (eBook)
DOI 10.1007/978-3-642-39386-0
Springer Heidelberg New York Dordrecht London

Printed on acid-free paper

Springer is part of Springer Science+Business Media (www.springer.com)

Dedicated to Jana and Jörn

Preface

Usually, the subject of a book on numerical mathematics pertains to a certain field of application or a certain numerical method. In this book we proceed in a different way. Stability is a concept that appears in various fields of numerical mathematics (as well as in other parts of mathematics). Although in each subfield stability is defined differently, there is a common meaning for this term, roughly described by the fact that perturbations are not amplifying the result in a dangerous way. In examining different fields of numerical mathematics concerning stability, we have the opportunity to recall some facts from numerical analysis. However, numerical mathematics cannot control stability exclusively for its own purpose. It turns out that stability is an ambiguous term, which also has strong connections to analysis and functional analysis.

Although stability is an essential requirement for numerical methods, the particular stability conditions are often not as obvious as, e.g., consistency conditions. The book may lead the reader to a better understanding of this term.

This book is an extension of a lecture held in the summer semester of 2003 at the University of Kiel (Christian-Albrechts-Universität zu Kiel). The exposition is self-contained, and the necessary facts from numerical analysis and analysis are provided. Hence, the book is well suited, e.g., as material for seminars in numerical mathematics.

The author wishes to express his gratitude to the publisher Springer for its friendly cooperation. In particular, he thanks Ann Kostant, editorial consultant for Springer, for polishing the English.

Leipzig, October 2013   *Wolfgang Hackbusch*

Contents

List of Symbols

Symbols

$\|\cdot\|_2$	spectral norm of a matrix; cf. §2.4.2.2
$\|\cdot\|_2$	norm of functions in $L^2(\Omega)$
$\|\cdot\|_\infty$	maximum or supremum norm of vectors or functions; cf. §2.4.2.2
$\|\cdot\|_\infty$	row-sum norm of matrices; cf. §2.4.2.2, §5.5.4.1
$\|\cdot\|_X$	norm of the Banach space X
$\|\cdot\|_X^*$	dual norm; cf. §3.3.2
$\|\cdot\|_{Y\leftarrow X}$	operator norm; cf. (3.23)
$\langle\cdot,\cdot\rangle_{X^*\times X}, \langle\cdot,\cdot\rangle_{X\times X^*}$	dual pairing; cf. §3.3.2

Greek Letters

$\delta a, \delta A, \ldots$	perturbations of quantities $a, A, \ldots$
$\Delta t, \Delta x$	step size in time and space; cf. §6.3.1
ε	often, symbol for an error
$\varepsilon_{\mathrm{abs}}$	absolute error; cf. footnote 1 on page 6
$\varepsilon_{\mathrm{rel}}$	relative error; cf. footnote 1 on page 6
$\eta(x,h), \eta_i$	discrete solution of ordinary differential equation at x or x_i; cf. §5.1.2
λ	often, an eigenvalue
λ	either $\Delta t/\Delta x$ or $\Delta t/\Delta x^2$ in §6; cf. (6.11)
λ	parameter of the Fredholm integral operator of the second kind; cf. (8.1)
Π_n	projection in §8.1.2.2
$\rho(\cdot)$	spectral radius; cf. (6.25)
$\sigma(M)$	spectrum of matrix M; cf. §5.5.4.1
$\tau(x, y(x); h)$	consistency error of a one-step method; cf. (5.18)
$\Phi_{i,n}(\cdot)$	Lagrange function; cf. (4.2)

$\phi(x_i, \eta_i, [\eta_{i+1},]h; f)$ function defining an explicit [implicit] one-step method; cf.
   §5.1.2, §5.4.1
$\phi(x_j, \eta_{j+r-1}, \ldots, \eta_j, h; f)$ function defining a multistep method; cf. (5.20a)
$\psi(\zeta)$ characteristic polynomial of a multistep method; cf. (5.21a)
Ω_n grid for difference method; cf. §7.2

Latin Letters

$a_{i,n}$	quadrature weights in §3
B	Banach space in §6
$\mathrm{cond}(\cdot)$	condition of a matrix; cf. (8.10)
$\mathbb{C}$	set of complex numbers
$C(\lambda, \Delta t)$	difference operator in §6.3.3
$C(D)$	space of continuous functions defined on D
$C_0(D)$	space of continuous functions with zero boundary value on ∂D
$C^k(D)$	$k \in \mathbb{N}_0$, space of functions with continuous derivatives of order $\leq k$
$C_0^\infty(\mathbb{R})$	infinitely differentiable functions with compact support
$C^\lambda(D)$	$\lambda > 0, \lambda \notin \mathbb{N}$, Hölder space; cf. §4.9
C_n	stability constant; cf. (3.13b), §4.3, §4.5
C_{stab}	stability constant; cf. (3.14), (4.5), §8.2.2, (7.11)
D	integration domain for the integral operator; cf. (8.2)
$\mathrm{degree}(\cdot)$	degree of a polynomial
E_j	shift operator; cf. (6.22)
eps	machine precision; cf. §2.3
$\mathcal{F}$	Fourier map; cf. §6.5.2
$\mathbf{f}_n$	right-hand side in finite difference equation (7.4)
$\mathrm{fl}(\cdot)$	operation in the argument performed in floating point arithmetic; cf. (2.13)
$G(\xi)$	characteristic function; cf. (6.34)
$G(x, \xi)$	characteristic function frozen at x; cf. (6.48)
h	step size for ordinary or partial differential equations; cf. §5.1.2
$H^k(\Omega), H^t(\Omega), H_0^k(\Omega), H_0^t(\Omega)$	Sobolev spaces; cf. pages 148, 149
I_n	interpolation; cf. (4.4)
J_n	Bessel function; cf. (2.2)
$k(x, y)$	kernel of the integral operator; cf. (8.2)
$k_n(x, y)$	approximation of $k(x, y)$; cf. (8.5)
K	integral operator; cf. (8.2)
K_n	discrete integral operator; cf. (8.4a)
ℓ^2, ℓ^∞	Banach spaces of sequences in $\mathbb{Z}$; cf. §6.3.1
L	elliptic differential operator in §7
$\mathcal{L}(X, Y)$	set of linear and continuous mappings from X to Y
$L_{i,n}(\cdot)$	Lagrange polynomial; cf. (4.3)
$\mathbf{L}_n$	finite difference matrix corresponding to L; cf. (7.4)

$L_n(\cdot)$ Legendre polynomial; cf. §3.1.4
$\mathcal{M}$ set of machine numbers; cf. §2.3
$\mathbb{N}$ natural numbers $\{1, 2, \ldots\}$
$\mathbb{N}_0$ $\mathbb{N} \cup \{0\} = \{0, 1, 2, \ldots\}$
$\mathbb{N}', \mathbb{N}''$ suitable infinite subsets of $\mathbb{N}$; cf. §7.2
p prolongation in §6.3.2
P often, polynomial
P_Y^n prolongation in §7.3.1
$Q_n(\cdot)$ quadrature rule; cf. §3
r restriction in §6.3.2
$r(\cdot)$ numerical radius; cf. §6.5.5
$\mathrm{rd}(\cdot)$ rounding to next machine number; cf. §2.3
$\mathbb{R}$ set of real numbers
R_X^n, R_Y^n restrictions in §7.3.1
$T(\cdot, h)$ trapezoidal rule of step size h; cf. §3.4.4
$T(t)$ solution operator; cf. §6.2
$\mathbf{u}_n$ solution of finite difference equation (7.4)
U_ν^μ discretisation of solution $u(t, x)$ in §6 at $t = \mu \Delta t$, $x = \nu \Delta x$
$x_{i,n}$ quadrature points in §3
X, Y often, Banach spaces
X, X_n, Y, Y_n Banach spaces in §7.3.1
X^* dual space $\mathcal{L}(X, \mathbb{R})$; cf. §3.3.2
$\hat{X}, \hat{X}_n$ Banach spaces in §7.3.2
$\mathbb{Z}$ entire numbers

Chapter 1
Introduction

In numerical mathematics we have to distinguish between two types of methods. There are finite algorithms, which solve a given task with a finite number of arithmetical operations. An example of such finite algorithms is the Gauss elimination for solving a system of linear equations. On the other hand, there are problems $\mathcal{P}$ that cannot be solved in finite time. Instead, there are (finite) approximation algorithms that involve solving substituting problems $\mathcal{P}_n$ producing results x_n, which, hopefully, tend to the true solution x of the original problem. The increasing closeness of $\mathcal{P}_n$ to $\mathcal{P}$ for $n \to \infty$ is the subject of the *consistency* condition. What really matters is the *convergence* $x_n \to x$. Whether consistency implies convergence depends on *stability*. It will turn out that under the assumption of consistency and possibly some technical conditions, convergence and stability are equivalent.

The original German manuscript has been used for Diplom students in mathematics. The material in this book is intended for master and Ph.D. students. Besides the discussion of the role of stability, a second goal is to review basic parts of numerical analysis.

We start in Chapter 2 with finite algorithms. We recall the condition of a problem and the stability of an algorithm. The amplification of input and intermediate floating point errors measures the quality of condition and stability. In this respect, the terms remain vague, since the amplification factors are some positive real numbers which may vary between 'small' (stable) and 'large' (unstable) without a clear separation.

Chapter 3 is devoted to quadrature methods, more precisely, to families of quadratures Q_n, where, with increasing n, the quality should improve ('consistency'). 'Stability' is again connected with the amplification of input errors. In contrast to Chapter 2, it is uniquely defined as to whether stability holds or not, since the terms 'small' and 'large' are replaced by finiteness or infiniteness of $\sup C_n$, the supremum of condition numbers C_n.

Although the stability definition is inspired by numerical phenomena, it is also suited to purely analytical purposes. Stability is almost equivalent to convergence of the quadrature result $Q_n(f)$ to the exact integral $\int f \mathrm{d}x$. Correspondingly,

W. Hackbusch, *The Concept of Stability in Numerical Mathematics*,
Springer Series in Computational Mathematics 45, DOI 10.1007/978-3-642-39386-0_1,
© Springer-Verlag Berlin Heidelberg 2014

analytical tools from functional analysis are involved, namely Weierstrass' approximation theorem and the uniform boundedness theorem.

Interpolation treated in Chapter 4 follows the same pattern as in Chapter 3. In both chapters one can pose the question of how important the stability statement $\sup C_n < \infty$ is, if one wants to perform only one quadrature or interpolation for a *fixed* n. In fact, polynomial interpolation is unstable, but when applied to functions of certain classes it behaves quite well.

This is different in Chapter 5, where one-step and multistep methods for the solution of ordinary initial-value problems are treated. Computing approximations requires an increasing number of steps, when the step size approaches zero. Often an instability leads to exponential growth of an error, eventually causing a termination due to overflow.

For ordinary differential equations instability occurs only for proper multistep methods, whereas one-step methods are always stable. This is different for partial differential equations, which are investigated in Chapter 6. Here, difference methods for hyperbolic and parabolic differential equations are treated. Stability describes the uniform boundedness of powers of the difference operators.

Also in the case of elliptic differential equations discussed in Chapter 7, stability is needed to prove convergence. In this context, stability describes the boundedness of the *inverse* of the difference operator or the finite element matrix independently of the step size.

The final chapter is devoted to Fredholm integral equations. Modern projection methods lead to a very easy proof of stability, consistency, and convergence. However, the Nyström method—the first discretisation method based on quadrature—requires a more involved analysis. We conclude the chapter with the analysis of the corresponding eigenvalue problem.

Despite the general concept of stability, there are different aspects to consider in the subfields. One aspect is the practical importance of stability (cf. §4.6), another concerns a possible conflict between a higher order of consistency and stability (cf. §3.5.2, Remark 4.15, Theorem 5.47, §6.6, §7.5.9).

Chapter 2
Stability of Finite Algorithms

2.1 About Algorithms

An algorithm is used to solve a (numerical) problem. For the mathematical formulation of a *problem* (or *task*) we use a mapping $\Phi : X \to Y$, which is to be evaluated numerically (cf. [9, Chap. 1]).

An algorithm is *executable* if the mapping Φ is composed of units that are realisable in a computer program. We call these units *elementary operations*. In the standard case, these elementary operations are the basic arithmetical operations $+, -, *, /$ in the set of real or whole numbers. In addition, the programming languages offer the use of some special functions like $\sin$, $\cos$, $\exp$, $\sqrt{\cdot}, \ldots$

An *algorithm* is the composition of elementary operations. A *finite algorithm* is characterised as involving only a finite number of elementary operations. The algorithm is a realisation of the mapping

$$\Phi : (x_1, \ldots, x_n) \in X = \text{domain}(\Phi) \mapsto (y_1, \ldots, y_m) \in Y = \text{range}(\Phi). \quad (2.1)$$

If Φ is realisable by at least one algorithm, then there are even infinitely many algorithms of this kind. Therefore, there is no one-to-one correspondence between a task and an algorithm.

A finite algorithm can be described by a sequence of vectors

$$\mathbf{x}^{(0)} = (x_1, \ldots, x_n), \; \mathbf{x}^{(1)}, \ldots, \mathbf{x}^{(j)} = (x_1^{(j)}, \ldots, x_{n_j}^{(j)}), \ldots, \mathbf{x}^{(p)} = (y_1, \ldots, y_m),$$

where the values $x_i^{(j)}$ from level j can be computed by elementary operations from the components of $\mathbf{x}^{(j-1)}$.

Example 2.1. *The scalar product* $y = \left\langle \left(\begin{smallmatrix} x_1 \\ x_2 \end{smallmatrix} \right), \left(\begin{smallmatrix} x_3 \\ x_4 \end{smallmatrix} \right) \right\rangle$ *has the input vector* $(x_1, \ldots, x_4)$. *The algorithm uses the intermediate values* $\mathbf{x}^{(1)} = (z_1, z_2)$ *with* $z_1 := x_1 x_2$, $z_2 := x_3 x_4$. *Then the output value is obtained from* $y := z_1 + z_2$.

The opposite of a finite algorithm is an infinite one, which, e.g., computes a sequence whose limit is the desired result of the problem. Since one has to terminate

W. Hackbusch, *The Concept of Stability in Numerical Mathematics*, Springer Series in Computational Mathematics 45, DOI 10.1007/978-3-642-39386-0_2, © Springer-Verlag Berlin Heidelberg 2014

such an infinite process, one finally obtains a finite algorithm producing an approx- imate result.

Here we only want to motivate the concept that algorithms should be constructed carefully regarding stability. It remains to analyse various concrete numerical methods (see, e.g., the monograph of Higham [5]).

2.2 A Paradoxical Example

2.2.1 First Algorithm

The following problem involves the family of *Bessel functions*. In such a case, one is well advised to look into a handbook of special functions. One learns that the n-th *Bessel function* (also called *cylinder function*) can be represented by a power series or as an integral:

$$J_n(x) = \sum_{k=0}^{\infty} \frac{(-1)^k (x/2)^{n+2k}}{k! (n+k)!} \qquad \text{for } n \in \mathbb{N}_0 \qquad (2.2)$$

$$= \frac{(-1)^n}{\pi} \int_0^{\pi} e^{\mathrm{i}x \cos(\varphi)} \cos(n\varphi) \mathrm{d}\varphi \qquad \text{for } n \in \mathbb{Z}. \qquad (2.3)$$

The chosen task is the computation of $J_5(0.6)$.

Assume that tabulated values of the first two Bessel functions J_0 and J_1 are offered as, e.g., in [13, page 99],

$$J_0(0.6) = 0.9120, \quad J_1(0.6) = 0.2867, \qquad (2.4)$$

but not the value of $J_5(0.6)$. Furthermore, assume that the book contains the recurrence relation

$$J_{n+1}(x) + J_{n-1}(x) = \frac{2n}{x} J_n(x) \qquad \text{for } n \in \mathbb{Z} \qquad (2.5)$$

as well as the property

$$\sum_{n=-\infty}^{\infty} J_n(x) = 1 \qquad \text{for all } x \in \mathbb{R}. \qquad (2.6)$$

Exercise 2.2. *Prove convergence of the series (2.2) for all $x \in \mathbb{C}$ (i.e., J_n is an entire function).*

An obvious algorithm solving our problem uses the recursion (2.5) for $n = 1, 2, 3, 4$ together with the initial values (2.4):

$$J_2(0.6) = -J_0(0.6) + \frac{2}{0.6}J_1(0.6) = -0.9120 + \frac{2}{0.6}0.2867 \qquad = 4.36667_{10}\text{-}2,$$
$$J_3(0.6) = -J_1(0.6) + \frac{4}{0.6}J_2(0.6) = -0.2867 + \frac{4}{0.6}4.36667_{10}\text{-}2 \qquad = 4.41111_{10}\text{-}3,$$
$$J_4(0.6) = -J_2(0.6) + \frac{6}{0.6}J_3(0.6) = -4.36667_{10}\text{-}2 + \frac{6}{0.6}4.41111_{10}\text{-}3 = 4.44444_{10}\text{-}4,$$
$$J_5(0.6) = -J_3(0.6) + \frac{8}{0.6}J_4(0.6) = -4.41111_{10}\text{-}3 + \frac{8}{0.6}4.44444_{10}\text{-}4 = 1.51481_{10}\text{-}3.$$

(2.7)

The result is obtained using only eight elementary operations. The underlying equations are exact. Nevertheless, the computed result for $J_5(0.6)$ is completely wrong, even the order of magnitude is incorrect! The exact result is $J_5(0.6) = 1.99482_{10}\text{-}5$.

Why does the computation fail? Are the tabulated values (2.4) misprinted? No, they are as correct as they can be. Is the (inexact) computer arithmetic, used in (2.5), responsible for the deviation? No, even exact arithmetic yields the same results. For those who are not acquainted with numerical effects, this might look like a paradox: exact computations using exact formulae yield completely wrong results.

2.2.2 Second Algorithm

Before we give an explanation, we show a second 'paradox': an algorithm based on inexact and even rather dubious formulae yields a perfect result. A numerical analyst asking for advice would recommend that we use the recurrence relation (2.5) in the opposite order; i.e., starting from $J_m(0.6)$ and $J_{m+1}(0.6)$ for some $m > 5$ and applying (2.5) in the order $n = m - 1, m - 2, \ldots$. The drawback is that neither $J_m(0.6)$ nor $J_{m+1}(0.6)$ are available. The expert's hint is to replace $J_{m+1}(0.6)$ by zero (vague reasoning: that value does not matter). The unknown value $J_m(0.6)$ will be obtained from the additional property (2.6).

We denote the candidates for $J_n(0.6)$ by j_n. The plan is to start from $j_{m+1} := 0$ and $j_m := J_m(0.6)$ and to apply (2.5): $j_{n-1} = \frac{2n}{0.6}j_n - j_{n+1}$. We observe that all results $j_{m-1}, j_{m-2}, \ldots$ depend linearly on j_m. Therefore, starting from

$$j'_{m+1} := 0, \qquad j'_m := 1 \tag{2.8}$$

and calculating

$$j'_{n-1} = \frac{2n}{0.6}j'_n - j'_{n+1} \qquad \text{for } n = m, m - 1, \ldots, 0, \tag{2.9}$$

we get quantities j'_n with the property $j_n = j_m j'_n$. Now the unknown value j_m can be determined from (2.6). From (2.2) or even from (2.5) we can derive that $J_{-n}(x) = (-1)^n J_n(x)$. Hence, (2.6) is equivalent to

$$J_0(x) + 2J_2(x) + 2J_4(x) + \ldots = 1. \tag{2.10}$$

Replacing the infinite sum by a finite one: $j_0 + 2\sum_{\nu=1}^{\lfloor m/2 \rfloor} j_{2\nu} = 1$, we arrive at $j_m \cdot \left(j'_0 + 2\sum_{\nu=1}^{\lfloor m/2 \rfloor} j'_{2\nu} \right) = 1$ and, in particular,

$$J_5(0.6) \approx j_m j_5' = j_5'/ \left(j_0' + 2 \sum_{\nu=1}^{\lfloor m/2 \rfloor} j_{2\nu}' \right).$$

The choice $m = 10$ yields

$$J_5(0.6) = 1.994819537430010\text{-}5,$$

where all 14 digits are correct.

Exercise 2.3. *Prove (2.10). Hint: comparison of the coefficients shows that (2.10) is equivalent to*

$$\sum_{m=0}^{\ell} \frac{(-1)^m}{(\ell - m)!\,(\ell + m)!} \times \left\{ \begin{array}{ll} 2, & \textit{if } m > 0 \\ 1, & \textit{if } m = 0 \end{array} \right\} = 0 \qquad \textit{for } \ell > 0.$$

For the latter identity use $e^x e^{-x} = 1$.

2.2.3 Explanation

As stated above, the tabulated values (2.4) are correct, since all four digits offered in the table are correct. It turns out that we have a fortunate situation since even the fifth digit is zero: the precise values with six digits are $J_0(0.6) = 0.912005$ and $J_1(0.6) = 0.286701$. However, the sixth and seventh decimals cause the *absolute errors*[1]

$$\varepsilon_{\text{abs}}^{(0)} = 0.9120049 - 0.9120 = 4.9_{10}\text{-}6, \quad \varepsilon_{\text{abs}}^{(1)} = 0.28670099 - 0.2867 = 9.9_{10}\text{-}7 \tag{2.11}$$

and the *relative errors*

$$\varepsilon_{\text{rel}}^{(0)} = \frac{4.9_{10}\text{-}6}{0.912005} = 5.3_{10}\text{-}6, \quad \varepsilon_{\text{rel}}^{(1)} = \frac{9.9_{10}\text{-}7}{0.286701} = 3.4_{10}\text{-}6.$$

Both are relatively small. This leads to the delusive hope that the same accuracy holds for $J_5(0.6)$. Instead, we observe the absolute and relative errors

$$\varepsilon_{\text{abs}}^{(5)} = 1.5_{10}\text{-}3, \quad \varepsilon_{\text{rel}}^{(5)} = 75. \tag{2.12}$$

As we see, the absolute error has increased (from about 10^{-6} in (2.11) to 10^{-3} in (2.12)). Additionally, the small value $J_5(0.6) \ll J_1(0.6)$ causes the large relative error in (2.12).

In order to understand the behaviour of the absolute error, we consider the recursion $j_{n+1} = \frac{2n}{0.6} j_n - j_{n-1}$ (cf. (2.5)) for general starting values j_0, j_1. Obviously,

---

[1] Let $\tilde{x}$ be any approximation of x. Then $\varepsilon_{\text{abs}} = x - \tilde{x}$ is the absolute error, while $\varepsilon_{\text{rel}} = \frac{x - \tilde{x}}{x}$ is the relative error.

all j_n depend linearly on j_0, j_1. In particular, we obtain

$$j_5 = \frac{74\,627}{27}\,j_1 - \frac{7820}{9}\,j_0.$$

One verifies that the valués $j_0 = 0.9120$ and $j_1 = 0.2867$ from (2.4) yield the value $j_5 = 1.51481_{10}\text{-}3$ from (2.7). The factor $\frac{74\,627}{27} \approx 2764$ is responsible for the strong amplification of the absolute error of j_1.

We add that Bessel functions are particular examples of linear recurrence relations that have been studied in the article of Oliver [6] (see also [3], [10]).

2.3 Accuracy of Elementary Operations

Floating point arithmetic used in computers is characterised by a fixed mantissa length. For t digits corresponding to the basis b ($t \geq 1, b \geq 2$), the set $\mathcal{M}$ of *machine numbers* consists of zero and all numbers of the form

$$x = \pm 0.d_1 \ldots d_t * b^E = \pm b^E \sum_{k=1}^{t} d_k b^{-k}, \text{ where}$$

$$1 \leq d_1 < b \text{ and } 0 \leq d_k < b \text{ for } k > 1, \ E \in \mathbb{Z}.$$

For simplicity, we ignore the boundedness of the exponent E which may lead to overflow or underflow.

Exercise 2.4. *(a) What is the best bound ε in $\sup_{\xi \in \mathbb{R}} \min_{x \in \mathcal{M}} |x - \xi|/|x| \leq \varepsilon$?*
(b) What is ε in the special case of the dual basis $b = 2$?

The minimiser x of $\min_{x \in \mathcal{M}} |x - \xi|$ is the best 'rounding' of $\xi \in \mathbb{R}$. Denoting this mapping by rd, we get the following estimate with machine precision eps $= \varepsilon$ (ε from Exercise 2.4):

$$|x - \text{rd}(x)| \leq \text{eps}|x|.$$

In the following we assume that eps is chosen such that all elementary operations 'op' fulfil the following inequality:

$$|\text{fl}\,(a \text{ op } b) - (a \text{ op } b)| \leq \text{eps}\,|a \text{ op } b| \qquad \text{for all } a, b \in \mathcal{M}. \tag{2.13}$$

Here, fl($\ldots$) indicates the evaluation of the operations indicated by '$\ldots$' in the sense of floating-point computer arithmetic. A similar estimate is assumed for elementary operations with one argument (e.g., fl($\sqrt{a}$)). Furthermore, we assume eps $\ll 1$.

Note that (2.13) controls the *relative* error.

Remark 2.5. *Inequality (2.13) can be regarded as a 'consistency' property.*

More details about computer arithmetic can be found, e.g., in [7, §16] or [8, §2.5].

2.4 Error Amplification

The mathematical term 'amplification' will be used in a quite general way. If an error ε is changed into $c\varepsilon$, we say that ε is *amplified* by c, even if $c \in (0, 1)$. The standard case of interest is, of course, whether $c > 1$ or even $c \gg 1$ may occur.

2.4.1 Cancellation

The reason for the disastrous result for $J_5(0.6)$ in the first algorithm can be discussed by a single operation:

$$y = x_1 - x_2. \tag{2.14}$$

On the one hand, this operation is harmless, since the consistency (2.13) of the subtraction guarantees that the floating point result $\tilde{y} = \mathrm{fl}(x_1 - x_2)$ satisfies the estimate[2]

$$|\tilde{y} - y| \le \mathrm{eps}\,|y|.$$

On the other hand, this estimate holds only for $x_1, x_2 \in \mathcal{M}$. The more realistic problem occurs when replacing the exact difference $\eta = \xi_1 - \xi_2$ with the difference $x_1 - x_2$ of machine numbers:

$$\eta = \xi_1 - \xi_2, \qquad x_1 = \mathrm{rd}(\xi_1),\ x_2 = \mathrm{rd}(\xi_2), \qquad \tilde{y} = \mathrm{fl}(x_1 - x_2),$$

where ξ_1 and ξ_2 are general real numbers. The interesting error is

$$|\eta - \tilde{y}| \qquad \text{for} \quad \tilde{y} = \mathrm{fl}(x_1 - x_2).$$

The absolute error is bounded by

$$\begin{aligned}
|\eta - \tilde{y}| &\le |x_1 - x_2 - \mathrm{fl}(x_1 - x_2)| + |\Delta x_1| + |\Delta x_2| \\
&\le \mathrm{eps}\,(|x_1 - x_2| + |\Delta x_1| + |\Delta x_2|) \\
\text{with } \Delta x_i &:= \xi_i - x_i.
\end{aligned}$$

As long as $|\eta| \sim |x_i|$ ($i = 1, 2$), the situation is under control, since also the relative error has size eps. Dramatic cancellation appears when $|\eta| \ll |x_1| + |x_2|$. Then the relative error is $\lesssim \mathrm{eps}\frac{|x_1|+|x_2|}{|\eta|}$. In the worst case, $|\eta|$ is of the size $\mathrm{eps}\,(|x_1| + |x_2|)$, so that the relative error equals $\mathcal{O}(1)$.

Cancellation takes place for values ξ_1 and ξ_2 having the same sign and similar size. On the other hand, the sum of two non-negative numbers is always safe, since $|\eta| = |\xi_1 + \xi_2| \ge |x_i|$ and $\frac{|\eta - \tilde{y}|}{|\eta|} \le 3\mathrm{eps}$. The factor 3 corresponds to the three truncations $\xi_1 \mapsto x_1, \xi_2 \mapsto x_2, x_1 + x_2 \mapsto \mathrm{fl}(x_1 + x_2)$.

---

[2] If x_1 and x_2 are close, often the exact difference is a machine number so that $\tilde{y} = y$.

2.4.1.1 Linear (Differential) Error Analysis

Given the mapping $y = \Phi(x)$ in (2.1), we have to check how errors Δx_i of the input value $\tilde{x}_i = x_i + \Delta x_i$ affect the output $\tilde{y} = \Phi(\tilde{x})$.

Assume that Φ is continuously differentiable. The Taylor expansion of $\tilde{y}_j = \Phi_j(x_1, \ldots, x_i + \Delta x_i, \ldots, x_n)$ yields $\tilde{y}_j = \Phi_j(x_1, \ldots, x_n) + \frac{\partial \Phi_j}{\partial x_i} \Delta x_i + o(\Delta x_i)$. Hence,

$$\left| \frac{\partial \Phi_j(x_1, \ldots, x_n)}{\partial x_i} \right| \tag{2.15}$$

is the amplification factor of the input error Δx_i, provided that we consider the *absolute* errors. The amplification of the *relative* error equals

$$\left| \frac{\partial \Phi_j(x_1, \ldots, x_n)}{\partial x_i} \right| \cdot \frac{|x_i|}{|y_j|}. \tag{2.16}$$

Which error is of interest (absolute or relative) depends on the particular problem. Instead of individual components x_i and y_j, one may also consider certain norms $\|x\|$ and $\|y\|$.

2.4.1.2 Condition and Stability

A problem is called *well-conditioned* if the amplification of the input error is not much larger than 1. On the other hand, if the error is strongly increased, the problem is called *ill conditioned*. The maximal error amplification factor is called *condition* or *condition number*.

Given a problem Φ, let $\mathcal{A}$ be any algorithm realising Φ. Let the algorithm use the following intermediate results:

$$\mathbf{x}^{(0)} = (x_1, \ldots, x_n) \underset{\varphi_1}{\mapsto} \mathbf{x}^{(1)} \underset{\varphi_2}{\mapsto} \ldots \underset{\varphi_p}{\mapsto} \mathbf{x}^{(p)} = \{y_1, \ldots, y_m\}.$$

We regain the mapping Φ of the problem as the composition $\Phi = \varphi_p \circ \varphi_{p-1} \circ \ldots \circ \varphi_1$. The mappings

$$\Phi^{(q)} := \varphi_p \circ \varphi_{p-1} \circ \ldots \circ \varphi_{q+1} \qquad \text{for } 0 \le q \le p$$

describe the transfer from $\mathbf{x}^{(q)}$ to $\mathbf{x}^{(p)} = \{y_1, \ldots, y_m\}$ (for $q = p$, the empty product is $\Phi^{(p)} = id$). While $\Phi^{(0)} = \Phi$ depends only on the underlying problem, the mappings $\Phi^{(q)}$ for $1 \le q < p$ depend on the algorithm. The derivative $\partial \Phi_j^{(q)} / \partial x_i^{(q)}$ corresponds to (2.15). It is the amplification factor describing $\Delta y_j / \Delta x_i^{(q)}$, where $\Delta x_i^{(q)}$ is the error of $x_i^{(q)}$ and Δy_j is the induced error of y_j. Since, by definition, the intermediate result $x_i^{(q)}$ is obtained by an elementary operation, its relative error is controlled by (2.13).

Let κ be the condition number of problem Φ. If all amplification factors

$|\partial \Phi_j^{(q)}/\partial x_i^{(q)}|$ are at most of size κ, the algorithm is called *stable*. Otherwise, the algorithm is called *unstable*.

Note that the terms stability/instability are not related to the problem, but to the algorithm. Since there are many algorithms for a given problem, one algorithm for problem Φ may be unstable, while another one could be stable.

Furthermore, we emphasise the *relative* relation: if the condition κ is large, also the amplification factors $|\partial \Phi_j^{(q)}/\partial x_i^{(q)}|$ of a stable algorithm may be large. In the case of the example in §2.2, the algorithms are stable. The difficulty is the large condition number.

We summarise: If a problem Φ possesses a large condition number, the disastrous amplification of the input errors cannot be avoided[3] by any algorithm realising Φ. If the problem is well-conditioned, one has to take care to choose a stable algorithm.

On purpose, the definitions of condition and stability are vague[4]. It is not fixed whether we consider relative or absolute errors, single components or a norm of the error. Furthermore, it remains open as to what amplification factors are considered to be moderate, large, or very large.

2.4.2 Further Examples

2.4.2.1 Evaluation of the Exponential Function

We consider the simple (scalar) problem $\Phi(x) = \exp(x)$ for $x = \pm 20$ under the hypothetical assumption that exp does not belong to the elementary operations.

First, we consider the condition. Here, it is natural to ask for the relative errors. According to (2.16), the condition number equals

$$\frac{\mathrm{d}\exp(x)}{\mathrm{d}x} \cdot \frac{|x|}{|\exp(x)|} = |x|.$$

Hence, moderate-sized values of x (like $x = \pm 20$) lead to a well-conditioned problem.

Any finite algorithm can only compute an approximate value of $\exp(x)$. We prescribe an accuracy ε and ask for an approximation $\widetilde{\exp}(x)$ satisfying

$$|\exp(x) - \widetilde{\exp}(x)| \leq \varepsilon \exp(x) \quad \text{with } \varepsilon := 10^{-6}. \tag{2.17}$$

We fix $x = 20$ and choose a truncated Taylor expansion

$$E_n(x) = \sum_{\nu=0}^{n} \frac{x^\nu}{\nu!}.$$

---

[3] The only exception occurs when the input values are exact machine numbers, i.e., $\Delta x_i = 0$.

[4] A systematic definition is attempted by de Jong [2].

The obvious algorithm for the evaluation of E_n is Horner's scheme consisting of a sequence of elementary operators. We use the fact that $n = 46$ leads to a remainder satisfying (2.17). Indeed, the computed value $E_n(20) = 4.851\,651_{10}8$ shows six correct leading digits, since the exact result is $\exp(20) = 4.851\,652_{10}8$.

Next, we consider $x = -20$. Obviously, the remainder $\left|\sum_{\nu=n-1}^{\infty} \frac{x^\nu}{\nu!}\right|$ for $x = -20$ is smaller than for $+20$. However, because of the relative error, the factor $\exp(x)$ on the right-hand side of (2.17) changes from $\exp(20)$ to $\exp(-20)$; hence, a more accurate approximation is required. One can check that $n = 79$ is sufficient. However, the computation of $E_n(-20)$ yields the value $3.9_{10}\text{-}9$, which deviates strongly from the exact value $2.061_{10}\text{-}9$. Obviously, this algorithm is not stable.

The reason for the latter failure is the cancellation of the positive terms in $\sum_{\nu=0}^{39} \frac{20^{2\nu}}{(2\nu)!} = 2.4258_{10}8$ with the negative ones in $\sum_{\nu=0}^{39} \frac{-20^{2\nu+1}}{(2\nu+1)!} = -2.4248_{10}8$.

A stable algorithm for approximating $\exp(-20)$ can be derived from the formula $\exp(-20) = 1/\exp(20) \approx 1/E_n(20)$. The choice $n = 45$ suffices to compute an approximation satisfying (2.17).

This example elucidates that analytical estimates of approximations as in (2.17) do not ensure at all that the evaluation of the approximating expression leads to acceptable results.

2.4.2.2 Zeros of Polynomials

The following example shows that standard concepts, e.g., from linear algebra, may lead to a completely unstable algorithm. We consider the eigenvalue problem for symmetric matrices (also here an approximation is unavoidable, since, in general, eigenvalues are not computable by a finite algorithm). Computing the eigenvalue of symmetric matrices is well-conditioned, as the following theorem will show. Since the notation of the spectral norm is needed in the theorem, we recall some facts about matrix norms. Given some vectors norms $\|\cdot\|$, the *associated matrix norm* is defined by

$$\|A\| := \max\{\|Ax\| / \|x\| : x \neq 0\}$$

(cf. (3.23) for the operator case). If $\|\cdot\| = \|\cdot\|_\infty$ is the maximum norm, the associated matrix norm is denoted by the identical symbol $\|\cdot\|_\infty$ and called the *row-sum norm* (cf. §5.5.4.1). The choice $\|\cdot\| = \|\cdot\|_2$ of the Euclidean vector norm leads us to the *spectral norm* $\|\cdot\|_2$ of matrices. An explicit description of the spectral norm is $\|A\|_2 = \max\{\sqrt{\lambda} : \lambda \text{ eigenvalue of } A^T A\}$.

Theorem 2.6. *Let $A, \Delta A \in \mathbb{R}^{n \times n}$ be matrices with A being symmetric (or normal), and set $\tilde{A} := A + \Delta A$. Then, for any eigenvalue $\tilde{\lambda}$ of $\tilde{A}$, there is an eigenvalue λ of A such that $|\tilde{\lambda} - \lambda| \leq \|\Delta A\|_2$.*

Proof. A diagonalisation by a unitary transformation does not change the eigenvalues and norms. Therefore, without loss of generality, $A = \text{diag}\{\lambda_1, \ldots\}$ is assumed to be diagonal. Let $\tilde{x} \neq 0$ be an eigenvector of $\tilde{A}$ corresponding to $\tilde{\lambda}$.

The identity

$$(\tilde{\lambda}I - A)^{-1}\,\Delta A\,\tilde{x} = (\tilde{\lambda}I - A)^{-1}\,(\tilde{A} - A)\,\tilde{x} = (\tilde{\lambda}I - A)^{-1}\,(\tilde{\lambda}I - A)\,\tilde{x} = \tilde{x}$$

proves $\|(\tilde{\lambda}I - A)^{-1}\,\Delta A\|_2 \geq 1$ for the spectral norm. We continue the inequality by $1 \leq \|(\tilde{\lambda}I - A)^{-1}\,\Delta A\|_2 \leq \|(\tilde{\lambda}I - A)^{-1}\|_2\,\|\Delta A\|_2$. Since $A = \mathrm{diag}\{\lambda_1,\ldots\}$, the norm equals $\|(\tilde{\lambda}I - A)^{-1}\|_2 = 1/\min\limits_{1\leq i \leq n} |\tilde{\lambda} - \lambda_i|$. Altogether, there is an eigenvalue λ of A such that $|\tilde{\lambda} - \lambda| = \min\limits_{i} |\tilde{\lambda} - \lambda_i| = 1/\|(\tilde{\lambda}I - A)^{-1}\|_2 \leq \|\Delta A\|_2$.   □

Since in linear algebra eigenvalues are introduced via the characteristic polynomial, one might get the following idea:

(a) Determine the characteristic polynomial $P(x) = \sum_{k=0}^{n} a_k x^k$.

(b) Compute the eigenvalues as roots of $P : P(x) = 0$.

For simplicity we assume that the coefficients a_k of $P(x) = \det(xI - A)$ can be determined exactly. Then the second part remains to be investigated: How are the zeros of P effected by perturbations of a_k? Here, the following famous example of Wilkinson is very informative (cf. [11] or [12, p. 54ff]).

We prescribe the eigenvalues (roots) $1, 2, \ldots, 20$. They are the zeros of

$$P(x) = \prod_{k=1}^{20} (i - x) = a_0 + \ldots + a_{19}x^{19} + a_{20}x^{20}$$

($a_0 = 20! = 2\,432\,902\,008\,176\,640\,000, \ldots, a_{19} = 190, a_{20} = 1$). The large value of a_0 shows the danger of an overflow during computations with polynomials. The determination of zeros of P seems to be rather easy, because P has only simple zeros and these are clearly separated.

We perturb only the coefficient a_{19} into $\tilde{a}_{19} = a_{19} - 2^{-23}$ ($2^{-23} = 1.192_{10}\text{-}7$) corresponds to 'single precision'). The zeros of the perturbed polynomial $\tilde{P}$ are

1,   2,   3,   4,   4.999 999 928,   6.000 006 9,   6.999 69,   8.0072,   8.917,

$10.09 \pm 0.64i$, $11.79 \pm 1.65i$, $13.99 \pm 2.5i$, $16.73 \pm 2.8i$, $19.5 \pm 1.9i$,   20.84.

The (absolute and relative) errors are not only of size $O(1)$, the appearance of five complex pairs shows that even the structure of the real spectrum is destroyed.

For sufficiently small perturbations, conjugate complex pairs of zeros cannot occur. Considering the perturbation $\tilde{a}_{19} = a_{19} - 2^{-55}$ ($2^{-55} = 2.776_{10}\text{-}17$), we obtain the zeros

$$1,\quad 2,\ldots,\quad 9,\quad 10,\quad 11 - 10^{-10},\quad 12 + 6_{10}\text{-}9,$$
$$13 - 17_{10}\text{-}9,\quad 14 + 37_{10}\text{-}9,\quad 15 - 59_{10}\text{-}9,\quad 16 + 47_{10}\text{-}9,\quad \ldots$$

For instance, the perturbation of the 15-th zero shows an error amplification by $59_{10}\text{-}9/2^{-55} = 2.1_{10}9$.

Stable algorithms for computing the eigenvalue directly are using the entries of the matrix and avoid the detour via the characteristic polynomial (at least in the form of $P(x) = \sum_{k=0}^{n} a_k x^k$; cf. Quarteroni et al. [8, §5]). In general, one should avoid polynomials in the classical representation as sums of monomials.

2.4.2.3 Numerical Differentiation and Related Phenomena

The numerical computation of (partial) derivatives is often required if the derivative is not implemented or if its evaluation is too costly. Assume $f \in C^3(I)$. The Taylor expansion

$$D_h f(x) := \frac{f(x+h) - f(x-h)}{2h} = f'(x) + \frac{h^2}{6} f'''(\xi)$$

with some $x - h < \xi < x + h$ yields the estimate $C_3 h^2$ of the error, where C_3 is a bound of f'''. However, for $h \to 0$ the truncation error spoils the computation because of the cancellation error. A very optimistic assumption is that the numerical evaluation of the function yields values $\tilde{f}(x \pm h)$ with the relative accuracy

$$\left| \tilde{f}(x \pm h) - f(x \pm h) \right| \leq |f(x \pm h)| \, \mathrm{eps} \leq C_0 \mathrm{eps},$$

where C_0 is a bound of f. Therefore, the difference quotient $D_h f$ is perturbed by an error bounded by $C_0 \mathrm{eps}/(2h)$, and the total error is

$$\varepsilon_h := C_3 h^2 + C_0 \mathrm{eps}/(2h).$$

Minimisation of this bound yields

$$h_{\mathrm{opt}} = \sqrt[3]{\frac{C_0}{4C_3}} \mathrm{eps} \quad \text{and} \quad \varepsilon_{h_{\mathrm{opt}}} = \frac{3}{4} 2^{\frac{2}{3}} C_0^{\frac{2}{3}} C_3^{\frac{1}{3}} \mathrm{eps}^{\frac{2}{3}}.$$

If f is of reduced smoothness, say $f \in C^\alpha(I)$, $1 < \alpha < 3$, the factor $\mathrm{eps}^{2/3}$ becomes $\mathrm{eps}^{1-\frac{1}{\alpha}}$, which is very poor for $\alpha > 1$ close to 1.

A similar phenomenon can appear in quite another situation. The following example is due to Bini–Lotti–Romani [1]. Let

$$f : \mathbb{R}^2 \times \mathbb{R}^2 \to \mathbb{R}^2$$

be the bilinear function

$$\begin{aligned} f_1(x, y) &= x_1 y_1, \\ f_2(x, y) &= x_2 y_1 + x_1 y_2. \end{aligned} \qquad (2.18)$$

We want to evaluate $f(x, y)$ with a minimal number of multiplications.[5] The standard implementation requires three multiplications $x_1 \cdot y_1$, $x_2 \cdot y_1$, and $x_1 \cdot y_2$.

---

[5] If we replace the scalars x_1, x_2, y_1, y_2 by matrices, the matrix-matrix multiplication is the dominant operation concerning cost.

Instead one can compute $f_\varepsilon : \mathbb{R}^2 \times \mathbb{R}^2 \to \mathbb{R}^2$ defined by

$$f_{\varepsilon,1}(x, y) = x_1 y_1 + \varepsilon^2 x_2 y_2,$$
$$f_{\varepsilon,2}(x, y) = x_2 y_1 + x_1 y_2$$

for some $\varepsilon > 0$. Obviously, for ε small enough, we obtain results for any prescribed accuracy. The second function can be evaluated by the algorithm

$$(x, y) \mapsto \begin{bmatrix} s_1 := (x_1 + \varepsilon x_2) \cdot (y_1 + \varepsilon y_2) \\ s_2 := (x_1 - \varepsilon x_2) \cdot (y_1 - \varepsilon y_2) \end{bmatrix} \mapsto f_\varepsilon := \begin{bmatrix} (s_1 + s_2)/2 \\ (s_1 - s_2)/(2\varepsilon) \end{bmatrix},$$

which requires only two multiplications. The expression $(s_1 - s_2)/(2\varepsilon)$ shows the analogy to the numerical differentiation, so that again cancellation occurs.

In fact, the numbers of multiplications (3 and 2, respectively) have their origin in tensor properties. We may rewrite (2.18) as

$$f_i = \sum_{1 \le j,k \le 2} v_{ijk} x_j y_k \qquad \text{for } i = 1, 2.$$

The coefficients are

$$v_{111} = v_{221} = v_{212} = 1 \text{ and } v_{ijk} = 0 \text{ otherwise.}$$

$\mathbf{v} = (v_{ijk})$ is an element of the tensor space $\mathbb{R}^2 \otimes \mathbb{R}^2 \otimes \mathbb{R}^2$. Its tensor rank is $\text{rank}(\mathbf{v}) = 3$; i.e., $r = 3$ is the minimal number of terms such that[6]

$$v_{ijk} = \sum_{\mu=1}^{r} a_i^{(\mu)} b_j^{(\mu)} c_k^{(\mu)}$$

for some vectors $a^{(\mu)}, b^{(\mu)}, c^{(\mu)} \in \mathbb{R}^2$ (cf. [4, Definition 3.32]). The solution for $r = 3$ is

$$a^{(1)} = b^{(1)} = b^{(3)} = c^{(1)} = c^{(2)} = \begin{bmatrix} 1 \\ 0 \end{bmatrix},$$
$$a^{(2)} = a^{(3)} = b^{(2)} = c^{(3)} = \begin{bmatrix} 0 \\ 1 \end{bmatrix}.$$

The corresponding coefficients $\mathbf{v}_\varepsilon = (v_{\varepsilon,ijk})$ of f_ε have tensor rank 2:

$$a^{(1)} = \begin{bmatrix} 1/2 \\ 1/(2\varepsilon) \end{bmatrix}, \qquad b^{(1)} = c^{(1)} = \begin{bmatrix} 1 \\ \varepsilon \end{bmatrix},$$
$$a^{(2)} = \begin{bmatrix} 1/2 \\ -1/(2\varepsilon) \end{bmatrix}, \qquad b^{(2)} = c^{(2)} = \begin{bmatrix} 1 \\ -\varepsilon \end{bmatrix}.$$

It can be shown that, whenever $\text{rank}(\mathbf{v}_\varepsilon) < \text{rank}(\mathbf{v})$ and $\mathbf{v}_\varepsilon \to \mathbf{v}$ as $\varepsilon \to 0$, then the computation of $\mathbf{v}_\varepsilon$ is unstable (cf. Hackbusch [4, Proposition 9.16]).

---

[6] Note that the minimal r in $v_{ij} = \sum_{\mu=1}^{r} a_i^{(\mu)} b_j^{(\mu)}$ is the usual matrix rank of (v_{ij}).

References

1. Bini, D., Lotti, G., Romani, F.: Approximate solutions for the bilinear form computational problem. SIAM J. Comput. **9**, 692–697 (1980)
2. de Jong, L.S.: Towards a formal definition of numerical stability. Numer. Math. **28**, 211–219 (1977)
3. Deuflhard, P.: A summation technique for minimal solutions of linear homogeneous difference equations. Computing **18**, 1–13 (1977)
4. Hackbusch, W.: Tensor spaces and numerical tensor calculus, *Springer Series in Computational Mathematics*, Vol. 42. Springer, Berlin (2012)
5. Higham, N.J.: Accuracy and stability of numerical algorithms. SIAM, Philadelphia (1996)
6. Oliver, J.: The numerical solution of linear recurrence relations. Numer. Math. **11**, 349–360 (1968)
7. Plato, R.: Concise Numerical Mathematics. AMS, Providence (2003)
8. Quarteroni, A., Sacco, R., Saleri, F.: Numerical Mathematics, 2nd ed. Springer, Berlin (2007)
9. Stoer, J., Bulirsch, R.: Introduction to Numerical Analysis. North-Holland, Amsterdam (1980)
10. Van der Cruyssen, P.: A reformulation of Olver's algorithm for the numerical solution of second-order linear difference equations. Numer. Math. **32**, 159–166 (1979)
11. Wilkinson, J.H.: Rounding errors in algebraic processes. Prentice-Hall, Englewood Cliffs (1964). Reprinted by Dover Publications, New York, 1994
12. Wilkinson, J.H.: Rundungsfehler. Springer, Berlin (1969)
13. Zeidler, E. (ed.): Oxford Users' Guide to Mathematics. Oxford University Press, Oxford (2004)

Chapter 3
Quadrature

In this chapter some facts from interpolation are used. Therefore, the reader may first have a look at §4.1 of the next chapter, which is devoted to interpolation. We prefer to start with quadrature instead of interpolation, since a projection between function spaces (interpolation) is more involved than a functional (quadrature).

3.1 Setting of the Problem and Examples

3.1.1 Quadrature Formulae

Any systematic approximation of an integral is called *quadrature*. We start with a simple setting of the problem.

Problem 3.1. *Assume that $f \in C([0,1])$. Compute an approximate value of the integral*

$$\int_0^1 f(x)\mathrm{d}x.$$

Other intervals and additional weight functions can be treated, but these generalisations are uninteresting for our purpose.

For all $n \in \mathbb{N}_0$ we define a quadrature formula

$$Q_n(f) = \sum_{i=0}^n a_{i,n} f(x_{i,n}). \tag{3.1}$$

Here $a_{i,n}$ are the quadrature weights and $x_{i,n}$ are the (disjoint) quadrature points. A sequence

$$\{Q_n : n \in \mathbb{N}_0\}$$

yields a *family of quadrature formulae* (sometimes also called a *quadrature rule* meaning that this rule generates quadrature formulae for all n).

W. Hackbusch, *The Concept of Stability in Numerical Mathematics*,
Springer Series in Computational Mathematics 45, DOI 10.1007/978-3-642-39386-0_3,
© Springer-Verlag Berlin Heidelberg 2014

3.1.2 Interpolatory Quadrature

The usual way to derive (3.1) is the 'interpolatory quadrature' via a (family of) interpolation methods (see §4.1). Consider an interpolation

$$f(x) \approx f_n(x) := \sum_{i=0}^{n} f(x_{i,n}) \Phi_{i,n}(x)$$

with Lagrange functions $\Phi_{i,n}$ (i.e., $\Phi_{i,n}$ belongs to the desired function space and satisfies $\Phi_{i,n}(x_{j,n}) = \delta_{ij}$; cf. (4.2)). Integrating f_n instead of f, one obtains

$$\int_0^1 f(x)\mathrm{d}x \approx \int_0^1 f_n(x)\mathrm{d}x = \sum_{i=0}^{n} f(x_{i,n}) \underbrace{\int_0^1 \Phi_{i,n}(x)\mathrm{d}x}_{=:a_{i,n}}.$$

The standard choice is polynomial interpolation. In this case, $\Phi_{i,n} = L_{i,n}$ are the Lagrange polynomials

$$L_{i,n}(x) := \prod_{k \in \{1,\dots,n\}\setminus\{i\}} \frac{x - x_{k,n}}{x_{i,n} - x_{k,n}}. \tag{3.2}$$

The advantage of interpolatory quadrature rules is their exactness for certain functions.

Remark 3.2. *Let the interpolatory quadrature be based on interpolation involving the functions $\Phi_{i,n}$. Then all functions from $U_n := \mathrm{span}\{\Phi_{i,n} : 0 \le i \le n\}$ are integrated exactly: $Q_n(f) = \int_0^1 f(x)\mathrm{d}x$ for $f \in U_n$. In the case of polynomial interpolation, U_n is the space of polynomials of degree $\le n$.*

The latter statement does not exclude that even more functions (e.g., polynomials of higher order) are also integrated exactly.

The next exercise refers to the case of polynomials, where $\Phi_{i,n} = L_{i,n}$ are the Lagrange polynomials (3.2).

Exercise 3.3. *The definition $a_{i,n} = \int_0^1 L_{i,n}(x)\mathrm{d}x$ is less helpful for its computation. Show that instead one can obtain the values $a_{i,n}$ by solving the following system of linear equations:*

$$\sum_{i=0}^{n} a_{i,n} x_{i,n}^k = \frac{1}{k+1} \qquad \text{for } k = 0, 1, \dots, n.$$

Hint: Verify that $Q_n(x^k) = \int_0^1 x^k \mathrm{d}x$.

So far the quadrature points $x_{i,n}$ are not fixed. Their choice determines the (family of) quadrature formulae.

3.1.3 Newton–Cotes Quadrature

Choose equidistant quadrature points in $[0, 1]$:[1]

$$x_{i,n} = \frac{i}{n} \qquad \text{for } i = 0, 1, \ldots, n.$$

The Newton–Cotes quadrature[2] is the interpolatory quadrature based on the polynomial interpolation at the nodes $x_{i,n}$.

Exercise 3.4. *Prove the symmetry $a_{i,n} = a_{n-i,n}$ $(0 \leq i \leq n)$ of the Newton–Cotes quadrature weights.*

For later use we cite the following asymptotic statement about the Newton–Cotes weights from Ouspensky [6] (also mentioned in [3, p. 79]):

$$a_{i,n} = \frac{(-1)^{i-1} n!}{i! \, (n-i)! n^2 \log^2 n} \left(\frac{1}{i} + \frac{(-1)^n}{n-i} \right) \left(1 + \mathcal{O}\left(\frac{1}{\log n} \right) \right) \qquad (3.3)$$

for $1 \leq i \leq n - 1$.

3.1.4 Gauss Quadrature

The *Legendre polynomial L_n* is a polynomial of degree n with the orthogonality property $\int_{-1}^{1} L_n(x)p(x)\mathrm{d}x = 0$ for all polynomials p of degree $\leq n - 1$ (cf. [7, §10.1.2]). This defines L_n uniquely up to a scalar factor, which is irrelevant for our purpose. We transform L_n from $[-1, 1]$ onto the integration interval $[0, 1]$: $\hat{L}_n(t) = L_n(2t - 1)$. It is well known that L_n has n distinct roots ξ_i in $[-1, 1]$. Hence $\hat{L}_n$ has n distinct roots $x_i := (1 + \xi_i)/2$ in $[0, 1]$. In the sequel, we rename $\hat{L}_n$ by L_n and call x_i the roots of the (transformed) Legendre polynomial.

Let $x_{i,n}$ $(0 \leq i \leq n)$ be the zeros of the (transformed) Legendre polynomial L_{n+1} (not L_n!). The interpolatory quadrature with these quadrature points yields the *Gauss quadrature*.[3]

3.2 Consistency

For the interpolatory quadrature defined via polynomial interpolation one uses the following consistency definition.

---

[1] The family of Newton–Cotes formulae is defined for $n \in \mathbb{N}$ only, not for $n = 0$. Formally, we may add $Q_0(f) := f(0)$.

[2] See Newton's remarks in [11, pp. 73-74].

[3] The original publication of Gauss [4] is from 1814. Christoffel [2] generalises the method to integrals with a weight function. For a modern description, see Stroud–Secrest [10].

Definition 3.5. *A family $\{Q_n : n \in \mathbb{N}_0\}$ of quadratures is called* consistent *if there is a function $g : \mathbb{N}_0 \to \mathbb{N}$ with $g(n) \to \infty$ for $n \to \infty$, so that*

$$Q_n(P) = \int_0^1 P(x)\mathrm{d}x \qquad \text{for all polynomials } P \text{ with degree}(P) \leq g(n). \quad (3.4)$$

An immediate consequence is the next statement.

Corollary 3.6. *Let $\{Q_n : n \in \mathbb{N}_0\}$ be consistent. Then for any polynomial P we have*

$$\lim_{n \to \infty} Q_n(P) = \int_0^1 P(x)\mathrm{d}x.$$

Proof. Set $\gamma := \mathrm{degree}(P)$. Because of $g(n) \to \infty$ for $n \to \infty$, there is an n_0 with $g(n) \geq \gamma$ for all $n \geq n_0$. Hence, $Q_n(P) = \int_0^1 P(x)\mathrm{d}x$ for $n \geq n_0$ proves the assertion. □

The quadrature families mentioned above satisfy (3.4) with the following values of $g(n)$.

Proposition 3.7. *The maximal orders $g(n)$ are*

$$g(n) = \begin{cases} n & n \text{ odd} \\ n+1 & n \text{ even} \end{cases} \qquad \text{for the Newton–Cotes quadrature,}$$

$$g(n) = 2n + 1 \qquad \text{for the Gauss quadrature.}$$

Proof. (i) Let P be a polynomial of degree n. The polynomial interpolation P_n of P in $n+1$ points yields $P_n = P$. Hence, by definition, any interpolatory quadrature satisfies $Q_n(P) = \int_0^1 P_n(x)\mathrm{d}x = \int_0^1 P(x)\mathrm{d}x$, which proves $g(n) \geq n$.

(ii) For even n, any polynomial of degree $n + 1$ has the form

$$P_{n+1} = a(x - 1/2)^{n+1} + P_n,$$

where $a \in \mathbb{C}$ and P_n has degree n. This proves

$$Q_n(P_{n+1}) = aQ_n((x - 1/2)^{n+1}) + Q_n(P_n).$$

$Q_n(P_n) = 0$ follows from Part (i), while in the case of the Newton–Cotes quadrature $Q_n((x - 1/2)^{n+1}) = 0$ can be concluded from Exercise 3.4.

(iii) Let P be a polynomial of degree $2n + 1$ and consider the Gauss quadrature. The Euclidean algorithms allows us to divide P by the (transformed) Legendre polynomial L_{n+1}: $P = pL_{n+1} + q$. Both polynomials p and q are of degree $\leq n$. The integral $\int_0^1 p(x)L_{n+1}(x)\mathrm{d}x$ vanishes because of the orthogonality property of L_{n+1}. On the other hand, $Q_n(pL_{n+1}) = 0$ follows from the fact that L_{n+1} vanishes at the quadrature points. Hence, $\int_0^1 P(x)\mathrm{d}x = \int_0^1 q(x)\mathrm{d}x =_{(i)} Q_n(q) = Q_n(P)$ proves the assertion. □

Later, we shall formulate an alternative (more general) consistency condition (cf. §3.4.9).

We conclude this section with consistency considerations for sufficiently smooth functions known as the Peano kernel representation of the quadrature error. Below we use the maximum norm

$$\|\varphi\|_\infty = \max\{|\varphi(x)| : x \in [0,1]\}. \tag{3.5}$$

Remark 3.8 (Peano kernel). *Let Q_n have maximal consistency order g_n. For some $m \in \mathbb{N}$ with $m \le g_n$ suppose that $f \in C^{m+1}([0,1])$. Then the quadrature error equals*

$$\int_0^1 f(x)\mathrm{d}x - Q_n(f) = \int_0^1 \pi_m(x,y)f^{(m+1)}(y)\mathrm{d}y \quad (\pi_m \text{ defined in the proof}). \tag{3.6a}$$

The error $\varepsilon_n := \left| \int_0^1 f(x)\mathrm{d}x - Q_n(f) \right|$ is estimated by

$$\varepsilon_n \le \alpha_{1,m}\|f^{(m+1)}\|_\infty \quad \text{and} \quad \varepsilon_n \le \alpha_{2,m}\|f^{(m+1)}\|_{L^2([0,1])} \tag{3.6b}$$

$$\text{with } \alpha_{1,m} := \int_0^1 |\pi_m(x,y)|\mathrm{d}y, \quad \alpha_{2,m} := \sqrt{\int_0^1 |\pi_m(x,y)|^2\mathrm{d}y}.$$

If $\pi_m(x,y)$ does not change sign, the following error equality holds for a suitable intermediate value $\xi \in [0,1]$:

$$\int_0^1 f(x)\mathrm{d}x - Q_n(f) = \alpha_m f^{(m+1)}(\xi) \quad \text{with} \quad \alpha_m := \int_0^1 \pi_m(x,y)\mathrm{d}y. \tag{3.6c}$$

Proof. The Taylor representation with remainder term yields $f(x) = P_m(x) + r(x)$, where the polynomial P_m of degree $\le m$ is irrelevant, since its quadrature error vanishes. Hence the quadrature error of f is equal to that of r. The explicit form of r is

$$r(x) = \frac{1}{m!}\int_0^x (x-y)^m f^{(m+1)}(y)\mathrm{d}y = \int_0^1 (x-y)_+^m f^{(m+1)}(y)\mathrm{d}y,$$

where $(t)_+ = t$ for $t \ge 0$ and $(t)_+ = 0$, otherwise. The quadrature error of r equals

$$\int_0^1 r(x)\mathrm{d}x - Q_n(r)$$

$$= \int_0^1 \int_0^1 (x-y)_+^m f^{(m+1)}(y)\mathrm{d}y\mathrm{d}x - \sum_{i=0}^n a_{i,n}\int_0^1 (x_{i,n}-y)_+^m f^{(m+1)}(y)\mathrm{d}y$$

$$= \int_0^1 \left[\int_0^1 (x-y)_+^m\mathrm{d}x - \sum_{i=0}^n a_{i,n}(x_{i,n}-y)_+^m \right]\int_0^1 f^{(m+1)}(y)\mathrm{d}y,$$

where the bracket defines the Peano kernel $\pi_m(x,y)$ in (3.6a). $\quad\square$

3.3 Convergence

3.3.1 Definitions and Estimates

One can pose different questions concerning convergence. Given a function f, we define the error

$$\varepsilon_n(f) := \left| Q_n(f) - \int_0^1 f(x)\mathrm{d}x \right|. \tag{3.7}$$

The first question concerns pure convergence: does $\lim_{n\to\infty} \varepsilon_n(f) = 0$ hold? For numerical purposes, $\lim_{n\to\infty} \varepsilon_n(f) = 0$ is a rather poor statement, since convergence may be arbitrarily slow. Usually, we have in mind a certain accuracy τ and try to perform the computation of $Q_n(f)$ for some n such that $\varepsilon_n(f) \le \tau$. Or, vice versa, we perform the computation for a *fixed* n and ask for an estimate of $\varepsilon_n(f)$ for this very n.

In the case of interpolatory quadrature, an estimate of $\varepsilon_n(f)$ for a fixed n can be obtained via the interpolation error $f - f_n$, since

$$\varepsilon_n(f) = \left| Q_n(f) - \int_0^1 f(x)\mathrm{d}x \right| = \left| \int_0^1 f_n(x)\mathrm{d}x - \int_0^1 f(x)\mathrm{d}x \right|$$

$$\le \int_0^1 |f_n(x) - f(x)|\,\mathrm{d}x \le \|f_n - f\|_\infty,$$

(for $\|\cdot\|_\infty$ compare §3.4.7.1). However, this estimate may be too pessimistic. An optimal error analysis can be based on the Peano kernel (cf. Remark 3.8). In any case one obtains bounds of the form

$$\varepsilon_n(f) \le c_n \|f^{(k_n)}\|_\infty, \tag{3.8}$$

involving derivatives of f of order $k_n \le g(n) + 1$.

Obviously, the right-hand side in (3.8) depends on c_n and f (including its derivatives). The quadrature family $\{Q_n\}$ is only responsible for the constants c_n, so that the convergence $c_n \|f^{(k_n)}\|_\infty \to 0$ cannot be taken as a property of $\{Q_n : n \in \mathbb{N}_0\}$. Moreover, the inequality (3.8) is applicable to $f \in C^k([0,1])$ only for n satisfying $k_n \le k$, which excludes $k_n \to \infty$.

One may ask for error estimates (3.8) with $k_n = 0$:

$$\varepsilon_n(f) \le c_n \|f\|_\infty \quad \text{(or more generally, } \varepsilon_n(f) \le c_n\|f^{(k)}\|_\infty \text{ with fixed } k\text{)}.$$

The answer is that in this case, $c_n \to 0$ cannot hold. For a proof, modify the constant function $f = 1$ in η-neighbourhoods of the quadrature points $x_{i,n}$ ($0 \le i \le n$) such that $0 \le \tilde{f} \le f$ and $\tilde{f}(x_{i,n}) = 0$. Because of $\tilde{f}(x_{i,n}) = 0$, we conclude that $Q_n(\tilde{f}) = 0$, while for sufficiently small η, the integral $\int_0^1 \tilde{f}(x)\mathrm{d}x$ is arbitrarily close to $\int_0^1 f(x)\mathrm{d}x = 1$ (the difference is bounded by $\delta := 2n\eta$). Since $\|f\|_\infty = \|\tilde{f}\|_\infty = 1$, we obtain no better error estimate than

$$\varepsilon_n(\tilde{f}) = \int_0^1 \tilde{f}(x)\mathrm{d}x \geq 1 - \delta = 1 \cdot \|\tilde{f}\|_\infty - \delta; \qquad \text{i.e., } c_n \geq 1.$$

This proves the following remark.

Remark 3.9. *Estimates of the form* $\varepsilon_n(f) \leq c_n \|f\|_\infty$ *require constants* $c_n \geq 1$.

Hence the right-hand sides $c_n \|f\|_\infty$ cannot be a zero sequence. Nevertheless, $\varepsilon_n(f) \to 0$ may hold. This leads to the next definition.

Definition 3.10. *A family* $\{Q_n : n \in \mathbb{N}_0\}$ *of quadrature formulae is called* convergent *if*

$$Q_n(f) \to \int_0^1 f(x)\mathrm{d}x \quad \text{for all } f \in C([0,1]). \tag{3.9}$$

Note that (3.9) is an equivalent formulation of $\varepsilon_n(f) \to 0$.

3.3.2 Functionals, Dual Norm, and Dual Space

Above we made use of the Banach space $X = C([0,1])$ equipped with the maximum norm $\|\cdot\|_X = \|\cdot\|_\infty$ from (3.5). The dual space of X consists of all linear and continuous mappings ϕ from X into $\mathbb{R}$. Continuity of ϕ is equivalent to boundedness; i.e., the following dual norm must be finite:

$$\begin{aligned}\|\phi\|_X^* &:= \sup\{|\phi(f)| : f \in X, \|f\|_X = 1\} \\ &= \sup\{|\phi(f)| / \|f\|_X : 0 \neq f \in X\}.\end{aligned}$$

Hence the dual space X^* is defined by

$$X^* := \{\phi : X \to \mathbb{R} \text{ linear with } \|\phi\|_X^* < \infty\}.$$

For the application of $\phi \in X^*$ onto an element $f \in X$, there are the following equivalent notations:

$$\phi(f) = \langle \phi, f \rangle_{X^* \times X} = \langle f, \phi \rangle_{X \times X^*}. \tag{3.10}$$

A consequence of the Hahn–Banach theorem (cf. Yosida [12, 1.§IV.4]) is the following statement.

Corollary 3.11. *For any* $f \in X$, *there is a functional* $\phi_f \in X^*$ *with*

$$\|\phi_f\|_X^* = 1 \quad \text{and} \quad \phi_f(f) = \|f\|_X.$$

In §2.4.1.2 we introduced the condition of a mapping by the amplification of perturbations. In the case of (linear) functionals there is a simple answer.

Remark 3.12. *A perturbation of $f \in X$ by $\delta f \in X$ is amplified by the mapping $\phi \in X^*$ by at most the factor $\|\phi\|_X^*$; i.e.,*

$$|\phi(f + \delta f) - \phi(f)| \le \|\phi\|_X^* \|\delta f\|_X .$$

Proof. Use $\phi(f + \delta f) - \phi(f) = \phi(\delta f)$ and the definition of $\|\phi\|_X^*$. □

Exercise 3.13. *The integral*

$$I(f) := \int_0^1 f(x)\mathrm{d}x$$

is a linear functional on $X = C([0, 1])$ with $\|I\|_X^* = 1$.

Another functional is the Dirac function(al) $\delta_a \in X^*$ defined by $\delta_a(f) := f(a)$. Here $\|\delta_a\|_X^* = 1$ holds.

The quadrature formula (3.1) is a functional which may be expressed in terms of Dirac functionals:

$$Q_n = \sum_{i=0}^{n} a_{i,n} \delta_{x_{i,n}}. \tag{3.11}$$

Lemma 3.14. *Let $X = C([0, 1])$. The quadrature formula (3.11) has the dual norm*

$$\|Q_n\|_X^* = \sum_{i=0}^{n} |a_{i,n}| .$$

Proof. Let $f \in X$ satisfy $\|f\|_X = 1$. Then

$$|Q_n(f)| = \left| \sum_{i=0}^{n} a_{i,n} f(x_{i,n}) \right| \le \sum_{i=0}^{n} |a_{i,n}|$$

holds because of $|f(x_{i,n})| \le \|f\|_X = 1$. This proves $\|Q_n\|_X^* \le \sum_{i=0}^{n} |a_{i,n}|$. The equality sign is obtained by choosing the following particular function $g \in X$. Set $g(x_{i,n}) := \mathrm{sign}(a_{i,n})$ and interpolate between the quadrature points and the end points 0, 1 linearly. Then g is continuous, i.e., $g \in X$, and $\|g\|_X = 1$. The definition yields the reverse inequality

$$\|Q_n\|_X^* = \sup\{|Q_n(f)| : f \in X, \|f\|_X = 1\} \ge |Q_n(g)| = \sum_{i=0}^{n} |a_{i,n}| . □$$

The convergence definition (3.9) now reads as $E_n(f) \to 0$ for all $f \in X$, where $E_n := Q_n - I$ is the error functional.

In principle, Q_n may consist of functionals other than point evaluations $\delta_{x_{i,n}}$. However, nonlocal functionals like $f \mapsto \int_0^1 q_{i,n}(x)f(x)\mathrm{d}x$ are impractical, since again they require an integration. One may take into account evaluations of derivatives: $f \mapsto f'(x_{i,n})$; e.g., interpolatory quadrature based on Hermite interpolation. In this case, the Banach space $X = C^1([0, 1])$ must be chosen.

3.4 Stability

3.4.1 Amplification of the Input Error

First, we introduce stability by a numerical argument. Let $f \in C([0,1])$ be the function to be integrated. The input data for $Q_n(f)$ are $\{f(x_{i,n}) : 0 \le i \le n\}$. Let $\tilde{f}$ be the floating point realisation[4] of f with absolute errors $\delta f_{i,n}$:

$$\tilde{f}_{i,n} := \tilde{f}(x_{i,n}) = f(x_{i,n}) + \delta f_{i,n} \qquad (0 \le i \le n).$$

It follows that

$$|\delta f_{i,n}| \le \|f - \tilde{f}\|_\infty.$$

First, we investigate the conditioning of integration and quadrature.

Remark 3.15. *(a) The integration problem $f \mapsto I(f) := \int_0^1 f(x)\mathrm{d}x$ is well-conditioned. The error estimate is*

$$|I(f) - I(\tilde{f})| \le \|f - \tilde{f}\|_\infty \qquad \text{for all } f, \tilde{f} \in C([0,1]). \tag{3.12}$$

(b) The corresponding error estimate for the quadrature Q_n is

$$|Q_n(\tilde{f}) - Q_n(f)| \le C_n \|f - \tilde{f}\|_\infty \tag{3.13a}$$

with the condition number $C_n := \|Q_n\|_X^* = \sum_{i=0}^{n} |a_{i,n}|.$ \tag{3.13b}

Proof. Use Remark 3.12, Exercise 3.13, and Lemma 3.14. □

3.4.2 Definition of Stability

The constant C_n from (3.13b) is the error amplification factor of the quadrature Q_n. To avoid an increasing error amplification, we have to require that C_n be uniformly bounded. This leads us directly to the stability definition.

Definition 3.16 (stability). *A quadrature family $\{Q_n : n \in \mathbb{N}_0\}$ is called stable if*

$$C_{\text{stab}} := \sup_{n \in \mathbb{N}_0} C_n = \sup_{n \in \mathbb{N}_0} \sum_{i=0}^{n} |a_{i,n}| < \infty. \tag{3.14}$$

An equivalent formulation is given below.

---

[4] To be quite precise, $\tilde{f}$ is only defined on the set $\mathcal{M}$ of machine numbers. However, for theoretical purposes, we may extend the function continuously (use, e.g., piecewise interpolation).

Remark 3.17. *(a) A family $\{Q_n : n \in \mathbb{N}_0\}$ of quadrature formulae is stable if and only if there is some C such that*

$$|Q_n(f)| \leq C\|f\|_\infty \qquad \text{for all } f \in C([0,1]) \text{ and all } n \in \mathbb{N}_0. \tag{3.15}$$

(b) C_{stab} from (3.14) is the minimal constant C in (3.15).

From Part (b) of the remark and after replacing f by $f - g$, one obtains the following estimate analogous to (3.13a).

Corollary 3.18. $|Q_n(f) - Q_n(g)| \leq C_{\text{stab}}\|f - g\|_\infty$ *holds for all $f, g \in C([0,1])$ and all $n \in \mathbb{N}_0$.*

Next we discuss the consequences of stability or instability in the numerical context. By $\tilde{f}$ we denoted the numerically evaluated function, which may be considered as a floating point version of f. The total error $Q_n(\tilde{f}) - \int_0^1 f(x)\mathrm{d}x$ can be estimated by the triangle inequality:

$$\left|Q_n(\tilde{f}) - \int_0^1 f(x)\mathrm{d}x\right| \leq \left|Q_n(\tilde{f}) - Q_n(f)\right| + \left|Q_n(f) - \int_0^1 f(x)\mathrm{d}x\right|$$
$$\leq C_n\|\tilde{f} - f\|_\infty + \varepsilon_n(f)$$

($\varepsilon_n(f)$ from (3.7)). If stability holds, the error is bounded by $C_{\text{stab}}\|\tilde{f}-f\|_\infty+\varepsilon_n(f)$ (cf. Corollary 3.18). Provided that $\varepsilon_n(f) \to 0$, the total error approaches the level of numerical noise $C_{\text{stab}}\|\tilde{f} - f\|_\infty$, which is unavoidable since it is caused by the input error $\|\tilde{f} - f\|_\infty$.

In the case of instability, $C_n \to \infty$ holds. While the term $\varepsilon_n(f)$ approaches zero, $C_n\|\tilde{f} - f\|_\infty$ tends to infinity. Hence, an enlargement of n does not guarantee a better result. If one does not have further information about the behaviour of $\varepsilon_n(f)$, it is difficult to find an n such that the total error is as small as possible.

In spite of what has been said about the negative consequences of instability, we have to state that the quality of a quadrature rule for *fixed* n has no relation to stability or instability. The sensitivity of Q_n to input errors is given only by the amplification factor C_n. Note that the size of C_n is not influenced by whether the sequence $(C_n)_{n\in\mathbb{N}}$ is divergent or convergent.

3.4.3 Stability of Particular Quadrature Formulae

Under the minimal condition that Q_n be exact for constants (polynomials of order zero; i.e., $g(n) \geq 0$ in (3.4)), one concludes that $1 = \int_0^1 \mathrm{d}x = Q_n(1) = \sum_{i=0}^n a_{i,n}$:

$$\sum_{i=0}^n a_{i,n} = 1. \tag{3.16}$$

Conclusion 3.19. *Assume (3.16). (a) If the quadrature weights are non-negative (i.e., $a_{i,n} \geq 0$ for all i and n), then the family $\{Q_n\}$ is stable with*

$$C_{\text{stab}} = C_n = 1 \qquad \text{for all } n.$$

(b) In general, the stability constant is bounded from below by $C_{\text{stab}} \geq 1$.

Proof. For (a) use $\sum_{i=0}^{n} |a_{i,n}| = \sum_{i=0}^{n} a_{i,n} = 1$. Part (b) follows from $\sum_{i=0}^{n} |a_{i,n}| \geq \left| \sum_{i=0}^{n} a_{i,n} \right| = |1| = 1$. □

The latter statement is based upon (3.16). A weaker formulation is provided next.

Exercise 3.20. *Conclusion 3.19 remains valid, if (3.16) is replaced by the condition $\lim_{n \to \infty} Q_n(1) = 1$.*

A very important property of the Gauss quadrature is the following well-known fact.

Lemma 3.21. *The Gauss quadrature has non-negative weights: $a_{i,n} \geq 0$. Hence the family of Gauss quadratures is stable with constant $C_{\text{stab}} = 1$.*

Proof. Define the polynomial $P_{2n}(x) := \prod_{0 \leq k \leq n, k \neq i} (x - x_{k,n})^2$ of degree $2n$. Obviously, $\int_0^1 P_{2n}(x)\mathrm{d}x > 0$. Since Q_n is exact for polynomials of degree $\leq 2n+1$, also $a_{i,n} P_{2n}(x_{i,n}) = Q_n(P_{2n}) > 0$. The assertion follows from $P_{2n}(x_{i,n}) > 0$. □

The Newton–Cotes formulae satisfy $a_{i,n} \geq 0$ only for $n \in \{1, 2, 3, 4, 5, 6, 7, 9\}$. On the other hand, the existence of some negative weights $a_{i,n} < 0$ does not necessarily imply instability. The following table shows the values of C_n from (3.13b):

n	1 to 7	8	9	10	11	12	14	16	18	20	22	24
C_n	1	1.45	1	3.065	1.589	7.532	20.34	58.46	175.5	544.2	1606	9923

Obviously, C_n increases exponentially to infinity; i.e., the Newton–Cotes formulae seem to be unstable. An exact proof of instability can be based on the asymptotic description (3.3) of $a_{i,n}$. For even n, the following inequality holds:

$$C_n = \sum_{i=0}^{n} |a_{i,n}| \geq \left| a_{\frac{n}{2},n} \right| \qquad (a_{i,n} \text{ from } (3.3)).$$

Exercise 3.22. *(a) Recall Stirling's formula for the asymptotic representation of $n!$ (cf. [13, §1.14.16], [5, Anhang 1]).*
(b) Using (a), study the behaviour of $\left| a_{\frac{n}{2},n} \right|$ and conclude that the family of Newton–Cotes formulae is unstable.

A further example of a quadrature rule follows in the next section.

3.4.4 Romberg Quadrature

The existence of negative weights $a_{i,n}$ is not yet a reason for instability, as long as $\sum_{i=0}^{n} |a_{i,n}|$ stays uniformly bounded. The following Romberg quadrature is an example of a stable quadrature involving negative weights.

For $h = 1/N$ with $N \in \mathbb{N}$, the sum

$$T(f,h) := h \left[\frac{1}{2} f(0) + f(h) + f(2h) + \ldots + f(1-h) + \frac{1}{2} f(1) \right]$$

represents the *compound trapezoidal rule*. Under the assumption $f \in C^m([0,1])$, m even, one can prove the *asymptotic expansion*

$$T(f,h) = \int_0^1 f(x)\mathrm{d}x + h^2 e_2(f) + \ldots + h^{m-2} e_{m-2}(f) + \mathcal{O}(h^m \|f^{(m)}\|_\infty) \quad (3.17)$$

(cf. Bulirsch [1], [7, §9.6]). Hence, the Richardson extrapolation is applicable: compute $T(f, h_i)$ for different h_i, $i = 0, \ldots, n$, and extrapolate the values

$$\left\{ \left(h_i^2, T(f, h_i) \right) : i = 0, \ldots, n \right\}$$

at $h = 0$. The result can be represented explicitly by the Lagrange polynomials (cf. Exercise 4.2):

$$Q_n(f) := \sum_{i=0}^{n} T(f, h_i) \underbrace{\prod_{\substack{\nu=0 \\ \nu \neq i}}^{n} \frac{h_\nu^2}{h_\nu^2 - h_i^2}}_{=:c_{i,n}} = \sum_{i=0}^{n} c_{i,n} T(f, h_i). \quad (3.18)$$

We fix an infinite sequence of step sizes

$$h_0 > h_1 > h_2 > \ldots, \qquad h_i = 1/N_i, \quad N_i \in \mathbb{N},$$

with the property

$$h_{i+1} \le \alpha h_i \qquad \text{with } 0 < \alpha < 1 \quad \text{for all } i \ge 0. \quad (3.19)$$

The original quadrature of Romberg [8] is based on $\alpha = 1/2$. Condition (3.19) enforces $h_i \to 0$ as $i \to \infty$. One infers from (3.19) that

$$h_i/h_j \le \alpha^{i-j} \text{ for } j \le i \qquad \text{and} \qquad h_i/h_j \ge \alpha^{i-j} \text{ for } j \ge i. \quad (3.20)$$

Lemma 3.23. *There is a constant $C < \infty$, so that $\sum_{i=0}^{n} |c_{i,n}| \le C$ for all $n \in \mathbb{N}_0$* ($c_{i,n}$ *from (3.18)*).

Proof. First, we recall simple inequalities, gathered in the next exercise.

Exercise 3.24. *Prove (a)* $1 + x \leq e^x$ *for all real x and (b)* $\frac{1}{1-x} \leq 1 + \vartheta x$ *with* $\vartheta = \frac{1}{1-x_0}$ *for all $0 \leq x \leq x_0 < 1$.*

(i) Part (b) with $\vartheta = \frac{1}{1-\alpha^2}$ yields

$$\prod_{j=1}^{m} \frac{1}{1 - \alpha^{2j}} \leq \prod_{j=1}^{m} \left(1 + \vartheta\alpha^{2j}\right) \leq \prod_{j=1}^{m} \exp\left(\vartheta\alpha^{2j}\right) \leq \prod_{j=1}^{\infty} \exp\left(\vartheta\alpha^{2j}\right) =: A,$$

where $A = \exp\left(\sum_{j=1}^{\infty} \vartheta\alpha^{2j}\right) = \exp\frac{\alpha^2\vartheta}{1-\alpha^2} = \exp\left(\alpha^2\vartheta^2\right)$. This implies that

$$\prod_{j=1}^{m} \frac{\alpha^{2j}}{1 - \alpha^{2j}} \leq A \prod_{j=1}^{m} \alpha^{2j} = A\alpha^{m(m+1)} \leq A\alpha^m \quad \text{for all } m \geq 0.$$

(ii) Split the product $\prod_{\nu \neq i}$ in (3.18) into the partial products $\prod_{\nu=0}^{i-1}$ and $\prod_{\nu=i+1}^{n}$. The first one is estimated by

$$\left| \prod_{\nu=0}^{i-1} \frac{h_\nu^2}{h_\nu^2 - h_i^2} \right| = \prod_{\nu=0}^{i-1} \frac{1}{1 - h_i^2/h_\nu^2} \underset{(3.20)}{\leq} \prod_{\nu=0}^{i-1} \frac{1}{1 - \alpha^{2(i-\nu)}} = \prod_{j=1}^{i} \frac{1}{1 - \alpha^{2j}} \underset{(i)}{\leq} A,$$

while the second one satisfies

$$\left| \prod_{\nu=i+1}^{n} \frac{h_\nu^2}{h_\nu^2 - h_i^2} \right| = \prod_{\nu=i+1}^{n} \frac{1}{h_i^2/h_\nu^2 - 1} \underset{(3.20)}{\leq} \prod_{\nu=i+1}^{n} \frac{1}{\alpha^{2(i-\nu)} - 1} = \prod_{j=1}^{n-i} \frac{1}{\alpha^{-2j} - 1}$$

$$= \prod_{j=1}^{n-i} \frac{\alpha^{2j}}{1 - \alpha^{2j}} \underset{(ii)}{\leq} A\alpha^{n-i}.$$

(iii) The estimate

$$\sum_{i=0}^{n} |c_{i,n}| = \sum_{i=0}^{n} \left| \prod_{\nu=0}^{i-1} \frac{h_\nu^2}{h_\nu^2 - h_i^2} \right| \times \left| \prod_{\nu=i+1}^{n} \frac{h_\nu^2}{h_\nu^2 - h_i^2} \right| \underset{(iii)}{\leq} A^2 \sum_{i=0}^{n} \alpha^{n-i}$$

$$< A^2 \sum_{j=0}^{\infty} \alpha^j = \frac{A^2}{1 - \alpha} =: C$$

proves the assertion. $\square$

The quadrature points $\{x_{j,n}\}$ used in Q_n are $\bigcup_{i=0}^{n}\{0, h_i, 2h_i, \ldots, 1\}$. For $h_i = 1/N_i$ such that N_i and N_j ($0 \leq i < j \leq n$) are relative prime,[5] the weight associated to the quadrature point $x_{j,n} = h_{n-1}$ is $a_{j,n} = \frac{1}{2}h_{n-1}c_{n-1,n}$. Because of $\text{sign}(c_{i,n}) = (-1)^{n-i}$, we conclude that Q_n contains negative weights.

---

[5] Under this assumption, the set of interior grid points $\{\nu h_i : 1 \leq \nu \leq N_i - 1\}$ are disjoint and the weights do not add up. However, even without this assumptions one finds grid points $x_{j,n}$ with negative weight.

Lemma 3.25. *The family of the Romberg quadratures $\{Q_n\}$ in (3.18) is stable.*

Proof. The compound trapezoidal rule $T(f, h_i) = \sum_{k=0}^{N_i} \tau_{k,i} f(kh_i)$ has the weights $\tau_{k,i} = h_i$ for $0 < k < N_i$ and $\tau_{k,i} = h_i/2$ for $k = 0, N_i$. In particular,

$$\sum_{k=0}^{N_i} |\tau_{k,i}| = \sum_{k=0}^{N_i} \tau_{k,i} = 1$$

holds. The quadrature formula Q_n is defined by

$$Q_n(f) = \sum_i c_{i,n} \sum_{k=0}^{N_i} \tau_{k,i} f(kh_i) = \sum_j a_{j,n} f(x_{j,n}) \quad \text{with } a_{j,n} := \sum_{(i,k):kh_i=x_{j,n}} c_{i,n} \tau_{k,i}.$$

Now, $\sum_j |a_{j,n}| \leq \sum_i |c_{i,n}| \sum_{k=0}^{N_i} |\tau_{k,i}| = \sum_i |c_{i,n}| \underset{\text{Lemma 3.23}}{\leq} C$ proves the stability condition. □

Lemma 3.26. *The family of the Romberg quadratures $\{Q_n\}$ in (3.18) is consistent.*

Proof. Let P be a polynomial of degree $\leq g(n) := 2n + 1$. In (3.17), the remainder term for $m := 2n + 2$ vanishes, since $P^{(m)} = 0$. This proves that $T(f, h)$ is a polynomial of degree $\leq n$ in the variable h^2. Extrapolation eliminates the terms $h_i^j e_j(P)$, $j = 2, 4, \ldots, 2n$, so that $Q_n(P) = \int_0^1 P(x) dx$. Since $g(n) = 2n + 1 \to \infty$ for $n \to \infty$, consistency according to Definition 3.5 is shown. □

The later Theorem 3.36 will prove convergence of the Romberg quadrature.

Exercise 3.27. *Condition $h_{i+1} \leq \alpha h_i$ from (3.19) can be weakened. Prove: if an $\ell \in \mathbb{N}$ and an $\alpha \in (0, 1)$ exist such that $h_{i+\ell} \leq \alpha h_i$ for all $i \geq 0$, then Lemma 3.23 remains valid.*

3.4.5 Approximation Theorem of Weierstrass

For the next step of the proof we need the well-known approximation theorem of Weierstrass.

Theorem 3.28. *For all $\varepsilon > 0$ and all $f \in C([0, 1])$ there is a polynomial $P = P_{\varepsilon,f}$ with $\|f - P\|_\infty \leq \varepsilon$.*

An equivalent formulation is: the set $\mathcal{P}$ of all polynomials is a *dense subset* of $C([0, 1])$.

In the following we prove a more general form (Stone–Weierstrass theorem). The next theorem uses the point-wise maximum $\text{Max}(f, g)(x) := \max\{f(x), g(x)\}$ and point-wise minimum $\text{Min}(f, g)(x) := \min\{f(x), g(x)\}$ of two functions. The following condition (i) describes that $\mathcal{F}$ is closed under these mappings. Condition (ii) characterises the approximability at two points ('separation of points').

Lemma 3.29. *Let $Q \subset \mathbb{R}^d$ be compact. Suppose that a family $\mathcal{F} \subset C(Q)$ of continuous functions satisfies the following two properties:*

(i) $\mathrm{Max}(f_1, f_2), \mathrm{Min}(f_1, f_2) \in \mathcal{F}$ *for all $f_1, f_2 \in \mathcal{F}$.*

(ii) For all $x', x'' \in Q$, all $\varepsilon > 0$, and all $g \in C(Q)$ there is a $\varphi \in \mathcal{F}$ such that $|\varphi(x') - g(x')| < \varepsilon$ *and* $|\varphi(x'') - g(x'')| < \varepsilon$.

Then for all $\varepsilon > 0$ and all $g \in C(Q)$ there exists a function $f \in \mathcal{F}$ with $\|f - g\|_\infty < \varepsilon$ (i.e., $\mathcal{F}$ is dense in $C(Q)$).

Proof. (a) Let $\varepsilon > 0$ and $g \in C(Q)$ be given. Fix some $x' = x_0 \in Q$, while the second point $x'' = y \in Q$ will be variable. By assumption (ii) there is a function $h = h(\cdot\,; x_0, y, \varepsilon)$ with

$$|h(x_0) - g(x_0)| < \varepsilon, \qquad |h(y) - g(y)| < \varepsilon.$$

The latter inequality yields in particular that $g(y) - \varepsilon < h(y)$. By continuity of h and g, this inequality holds in a whole neighbourhood $U(y)$ of y:

$$g(x) - \varepsilon < h(x) \qquad \text{for all } x \in U(y).$$

Since $\bigcup_{y \in Q} U(y)$ covers the compact set Q, there is a finite subset of neighbourhoods $\{U(y_i) : i = 1, \dots, n\}$ covering Q: $\bigcup_{i=1}^n U(y_i) \supset Q$. Each y_i is associated to a function $h(\cdot\,; x_0, y_i, \varepsilon) \in \mathcal{F}$ with

$$g(x) - \varepsilon < h(x; x_0, y_i, \varepsilon) \qquad \text{for all } x \in U(y_i).$$

By assumption (i), $h(\cdot\,; x, \varepsilon) := \mathrm{Max}_{i=1,\dots,n} h(\cdot\,; x_0, y_i, \varepsilon)$ again belongs to $\mathcal{F}$ and satisfies

$$g(x) - \varepsilon < h(x; x_0, \varepsilon) \qquad \text{for all } x \in Q.$$

(b) Next, the parameter x_0 becomes a variable in Q. Since all $h(\cdot\,; x_0, y_i, \varepsilon)$ approximate the function g at x_0, the opposite inequality $g(x_0) + \varepsilon > h(x_0; x_0, \varepsilon)$ holds. Again, there is a neighbourhood $V(x_0)$, so that

$$g(x) + \varepsilon > h(x; x_0, \varepsilon) \qquad \text{for all } x \in V(x_0).$$

As in Part (a), one finds a finite covering $\{V(x_i) : i = 1, \dots, m\}$ of Q. The function

$$f := \mathrm{Min}_{i=1,\dots,m} h(\cdot\,; x_i, \varepsilon)$$

belongs again to $\mathcal{F}$ and satisfies $g + \varepsilon > f$. Since each $h(\cdot\,; x_i, \varepsilon)$ satisfies the inequality $g - \varepsilon < h(\cdot\,; x_i, \varepsilon)$ from Part (a), also $g - \varepsilon < f$ follows. Together, one obtains $\|f - g\|_\infty < \varepsilon$; i.e., $f \in \mathcal{F}$ is the desired approximant. $\square$

Remark 3.30. *Instead of the lattice operations* Max *and* Min, *one can equivalently require that $\mathcal{F}$ be closed with respect to the absolute value:*

$$f \in \mathcal{F} \quad \Rightarrow \quad |f| \in \mathcal{F},$$

where $|f|$ is defined point-wise: $|f|(x) := |f(x)|$ for all $x \in Q$.

Proof. Use $\text{Max}(f, g) = \frac{1}{2}(f + g) + \frac{1}{2}|f - g|$ and $\text{Min}(f, g) = \frac{1}{2}(f + g) - \frac{1}{2}|f - g|$ and in the reverse direction $|f| = \text{Max}(f, -f)$. □

The addition and multiplication of functions is defined point-wise: $(f + g)(x) = f(x) + g(x)$, $(f \cdot g)(x) = f(x)g(x)$, $x \in Q$. Correspondingly, multiplication by scalars from the field $\mathbb{K}$ is defined point-wise: $(\lambda f)(x) = \lambda f(x)$.

Definition 3.31 (algebra of functions). *A set $\mathcal{A}$ of functions is called an algebra, if $\mathcal{A}$ (without multiplication) is a vector space, and, additionally, is equipped with the multiplication satisfying the usual distributive law.*

Example 3.32. *Examples of algebras are all (a) continuous functions on $Q \subset \mathbb{R}^d$ (no compactness of Q required), (b) bounded functions on $Q \subset \mathbb{R}^d$, (c) polynomials, (d) trigonometric functions.*

In the case of (d) in the previous example one has to show that, e.g., the product $\sin(nx)\cos(mx)$ $(n, m \in \mathbb{N}_0)$ is again a trigonometric function. This follows from $2\sin(nx)\cos(mx) = \sin((n + m)x) + \sin((n - m)x)$.

If $\mathcal{A} \subset C(Q)$ is an algebra, the closure $\bar{\mathcal{A}}$ (with respect to the maximum norm $\|\cdot\|_\infty$) is called the *closed hull of the algebra* $\mathcal{A}$.

Exercise 3.33. *If $\mathcal{A}$ is an algebra of continuous functions, also $\bar{\mathcal{A}}$ is an algebra of continuous functions; i.e., $f, g \in \bar{\mathcal{A}}$ implies $f + g \in \bar{\mathcal{A}}$ and $f \cdot g \in \bar{\mathcal{A}}$.*

Lemma 3.34 (Weierstrass). *Let $\mathcal{A} \subset C(Q)$ be an algebra. Then $|f| \in \bar{\mathcal{A}}$ for all $f \in \mathcal{A}$.*

Proof. (i) A simple scaling argument shows that it suffices to show the assertion for $f \in \mathcal{A}$ with $\|f\|_\infty \leq 1$.

(ii) Let $\varepsilon > 0$ be given. The function $T(\zeta) := \sqrt{\zeta + \varepsilon^2}$ is holomorphic in the complex half-plane $\Re \zeta > -\varepsilon^2$. The Taylor series $\sum \alpha_\nu (x - \frac{1}{2})^\nu$ of $T(x)$ has the convergence radius $\frac{1}{2} + \varepsilon^2$ and converges uniformly in the interval $[0, 1]$. Hence there is a finite Taylor polynomial P_n of degree n, so that

$$\left| \sqrt{x + \varepsilon^2} - P_n(x) \right| \leq \varepsilon \qquad \text{for all } 0 \leq x \leq 1.$$

(iii) Replacing x by x^2, we obtain

$$\left| \sqrt{x^2 + \varepsilon^2} - P_n(x^2) \right| \leq \varepsilon \qquad \text{for all } -1 \leq x \leq 1.$$

The particular case $x = 0$ shows $|\varepsilon - P_n(0)| \leq \varepsilon$ and, therefore, $|P_n(0)| \leq 2\varepsilon$. The polynomial $Q_{2n}(x) := P_n(x^2) - P_n(0)$ of degree $2n$ has a vanishing absolute term and satisfies

$$\left| \sqrt{x^2 + \varepsilon^2} - Q_{2n}(x) \right| \leq \left| \sqrt{x^2 + \varepsilon^2} - P_n(x^2) \right| + |P_n(0)| \leq 3\varepsilon$$

for all $-1 \leq x \leq 1$. Using

$$\left| \sqrt{x^2 + \varepsilon^2} - |x| \right| = \frac{\varepsilon^2}{\left| \sqrt{x^2 + \varepsilon^2} + |x| \right|} \leq \frac{\varepsilon^2}{\varepsilon} = \varepsilon \qquad \text{for all } -1 \leq x \leq 1,$$

we obtain the inequality

$$\left| |x| - Q_{2n}(x) \right| \leq 4\varepsilon \qquad \text{for all } -1 \leq x \leq 1.$$

(iv) For all f with $\|f\|_\infty \leq 1$, the values $f(\xi)$ ($\xi \in Q$) satisfy the inequality $-1 \leq f(\xi) \leq 1$, so that $x = f(\xi)$ can be inserted into the last inequality:

$$\left| |f(\xi)| - Q_{2n}(f(\xi)) \right| \leq 4\varepsilon \qquad \text{for all } \xi \in Q.$$

Because of[6]

$$Q_{2n}(f(\xi)) = \sum_{\nu=1}^{n} q_\nu \left(f(\xi) \right)^{2\nu} = \left(\sum_{\nu=1}^{n} q_\nu f^{2\nu} \right)(\xi),$$

$Q_{2n}(f)$ belongs again to $\mathcal{A}$ and satisfies the estimate $\||f| - Q_{2n}(f)\| \leq 4\varepsilon$. As $\varepsilon > 0$ is arbitrary, $|f|$ belongs to the closure of $\mathcal{A}$. □

Now we prove the theorem of Weierstrass in the generalised form of Stone:

Theorem 3.35 (Stone–Weierstrass). *Assume that*

(i) $Q \subset \mathbb{R}^d$ is compact,

(ii) $\mathcal{A}$ is an algebra of continuous functions on Q (i.e., $\mathcal{A} \subset C(Q)$),

(iii) $\mathcal{A}$ separates the points of Q; i.e., for any pair of points $x', x'' \in Q$ with $x' \neq x''$ there is a function $f \in \mathcal{A}$ with $f(x') \neq f(x'')$.

Then the closed hull $\bar{\mathcal{A}}$ satisfies either $\bar{\mathcal{A}} = C(Q)$ or there is an $x_0 \in Q$ so that

$$\bar{\mathcal{A}} = \{f \in C(Q) : f(x_0) = 0\}. \tag{3.21}$$

Proof. (a) By Lemma 3.34 and Remark 3.30, $\mathcal{F} = \bar{\mathcal{A}}$ satisfies the requirement (i) of Lemma 3.29. As soon as (ii) from Lemma 3.29 is shown, $\bar{\mathcal{F}} = C(Q)$ follows. Since $\mathcal{F} = \bar{\mathcal{A}}$ is already closed, the first case $\bar{\mathcal{A}} = C(Q)$ follows.

(b) For the proof of (ii) from Lemma 3.29, we consider the following alternative: *either* for any $x \in Q$ there is an $f \in \mathcal{A}$ with $f(x) \neq 0$ *or* there exists an $x_0 \in Q$ with $f(x_0) = 0$ for all $f \in \mathcal{A}$. The first alternative will be investigated in (c), the second one in (d).

(c) Assume the first alternative. First we prove the following:
Assertion: For points $x', x'' \in Q$ with $x' \neq x''$ from Assumption (ii) in Lemma 3.29 there exists an $f \in \mathcal{A}$ with $0 \neq f(x') \neq f(x'') \neq 0$.

For its proof we use the separability property (iii): $f(x') \neq f(x'')$ holds for a suitable f. Assume $f(x') = 0$, which implies $f(x'') \neq 0$ (the case $f(x'') = 0$ and

---

[6] Here we use that $q_0 = 0$, since $f^0 = 1$ may not necessarily belong to the algebra $\mathcal{A}$.

$f(x') \neq 0$ is completely analogous). The first alternative from Part (b) guarantees the existence of an $f_0 \in \mathcal{A}$ with $f_0(x') \neq 0$. The function

$$f_\lambda := f - \lambda f_0$$

has the properties

$$
\begin{array}{lll}
0 \neq f_\lambda(x'') & \text{for sufficiently small } \lambda, & (\text{because of } f(x'') \neq 0) \\
f_\lambda(x'') \neq f_\lambda(x') & \text{for sufficiently small } \lambda, & (\text{because of } f(x') \neq f(x'')) \\
f_\lambda(x') = \lambda f_0(x') \neq 0 & \text{for all } \lambda \neq 0. &
\end{array}
$$

Hence, for sufficiently small but positive λ, we have

$$0 \neq f_\lambda(x') \neq f_\lambda(x'') \neq 0, \qquad f_\lambda \in \mathcal{A},$$

and f_λ (renamed by f) has the required properties. This proves the assertion.

Let $g \in C(Q)$ be the function from assumption (ii) of Lemma 3.29. Concerning the required φ, we make the ansatz $\varphi = \alpha f + \beta f^2$ with f from the assertion. $f \in \mathcal{A}$ implies that also $f^2, \varphi \in \mathcal{A}$. The conditions $\varphi(x') = g(x')$ and $\varphi(x'') = g(x'')$ lead to a 2×2-system of linear equations for α and β. Since the determinant

$$f(x')f^2(x'') - f(x'')f^2(x') = f(x')f(x'')\,[f(x'') - f(x')]$$

does not vanish, solvability is ensured. Therefore, assumption (ii) of Lemma 3.29 is satisfied even for $\varepsilon = 0$ and Part (a) shows $\bar{\mathcal{A}} = C(Q)$.

(d) Assume the second alternative: there is an $x_0 \in Q$ with $f(x_0) = 0$ for all $f \in \mathcal{A}$. Let $\mathbf{1} \in C(Q)$ be the function with constant value 1. Denote the algebra generated from $\mathcal{A}$ and $\{\mathbf{1}\}$ by $\mathcal{A}^*$; i.e., $\mathcal{A}^* = \{f = g + \lambda \mathbf{1} : g \in \mathcal{A}, \lambda \in \mathbb{K}\}$. Obviously, for any $x \in Q$ there is an $f \in \mathcal{A}^*$ with $f(x) \neq 0$ (namely $f = \mathbf{1}$). Hence the first alternative applies to $\mathcal{A}^*$. The previous proof shows $\bar{\mathcal{A}}^* = C(Q)$.

Let $g \in C(Q)$ be an arbitrary function with $g(x_0) = 0$, i.e., belonging to the right-hand side of (3.21). Because of $g \in C(Q) = \bar{\mathcal{A}}^*$, for all $\varepsilon > 0$ there is an $f^* \in \mathcal{A}^*$ with $\|g - f^*\|_\infty < \varepsilon$. By definition of $\mathcal{A}^*$ one may write f^* as $f^* = f + \lambda \mathbf{1}$ with $f \in \mathcal{A}$. This shows that

$$\|g - f - \lambda \mathbf{1}\|_\infty < \varepsilon.$$

In particular, at x_0 we have $|g(x_0) - f(x_0) - \lambda| = |\lambda| < \varepsilon$. Together, one obtains $\|g - f\|_\infty < 2\varepsilon$, where $f \in \mathcal{A}$. This proves (3.21). $\square$

For the proof of Theorem 3.28 choose $Q = [0,1]$ (compact subset of $\mathbb{R}^1$) and $\mathcal{A}$ as the algebra of all polynomials. For this algebra, assumption (iii) of Theorem 3.35 is satisfied with $f(x) = x$. Hence, one of the two alternatives $\bar{\mathcal{A}} = C(Q)$ or (3.21) holds. Since the constant function $\mathbf{1}$ belongs to $\mathcal{A}$, (3.21) is excluded and $\bar{\mathcal{A}} = C(Q)$ is shown.

3.4.6 Convergence Theorem

We recall Definition 3.10: Q_n is convergent if $Q_n(f) \to \int_0^1 f(x)\mathrm{d}x$ holds for all $f \in C([0,1])$. So far it has remained open as to whether this property would hold. Note that all previous convergence results require more smoothness than continuity of f. Now we can give a positive answer. The next theorem follows the pattern

$$\text{consistency} + \text{stability} \implies \text{convergence.} \qquad (3.22)$$

Theorem 3.36 (convergence theorem). *If the family $\{Q_n : n \in \mathbb{N}_0\}$ of quadrature formulae is consistent and stable, then it is convergent.*

Proof. Let $\varepsilon > 0$ be given. We have to show that for all $f \in C([0,1])$ there is an n_0 such that

$$\left| Q_n(f) - \int_0^1 f(x)\mathrm{d}x \right| \leq \varepsilon \qquad \text{for } n \geq n_0.$$

Let P be an arbitrary polynomial. The triangle inequality yields

$$\left| Q_n(f) - \int_0^1 f(x)\mathrm{d}x \right|$$

$$= \left| Q_n(f) - Q_n(P) + Q_n(P) - \int_0^1 P(x)\mathrm{d}x + \int_0^1 P(x)\mathrm{d}x - \int_0^1 f(x)\mathrm{d}x \right|$$

$$\leq \left| Q_n(f) - Q_n(P) \right| + \left| Q_n(P) - \int_0^1 P(x)\mathrm{d}x \right| + \left| \int_0^1 P(x)\mathrm{d}x - \int_0^1 f(x)\mathrm{d}x \right|.$$

We choose P according to Theorem 3.28 such that $\|f - P\|_\infty \leq \varepsilon/(1 + C_{\text{stab}})$, where C_{stab} is the stability constant. Now, thanks to Corollary 3.18, the first term $|Q_n(f) - Q_n(P)|$ can be estimated by

$$C_{\text{stab}} \|f - P\|_\infty \leq \varepsilon C_{\text{stab}}/(1 + C_{\text{stab}}).$$

The chosen P has a fixed degree(P). Because of $g(n) \to \infty$ there is an n_0 such that $g(n) \geq \text{degree}(P)$ for all $n \geq n_0$. Hence, consistency guarantees exactness of the quadrature: $\left| Q_n(P) - \int_0^1 P(x)\mathrm{d}x \right| = 0$.
Remark 3.15 yields

$$\left| \int_0^1 P(x)\mathrm{d}x - \int_0^1 f(x)\mathrm{d}x \right| \leq \|f - P\|_\infty \leq \varepsilon/(1 + C_{\text{stab}})$$

for the last term.
Together, the sum of the three terms is bounded by $\frac{\varepsilon C_{\text{stab}}}{1 + C_{\text{stab}}} + \frac{\varepsilon}{1 + C_{\text{stab}}} = \varepsilon$. $\quad\square$

According to Theorem 3.36, stability is *sufficient* for convergence. Next we show that the stability condition (3.14) is also *necessary* for convergence. As a tool, we need a further theorem from functional analysis.

3.4.7 Uniform Boundedness Theorem

3.4.7.1 Banach Space Notations

X is called a *normed (linear) space* (the norm may be expressed by the explicit notation $(X, \|\cdot\|)$), if a norm $\|\cdot\|$ is defined on the vector space X. If necessary, we write $\|\cdot\|_X$ for the norm on X.

X is called a *Banach space*, if X is normed and complete ('complete' means that all Cauchy sequences converge in X).

By $\mathcal{L}(X, Y)$ we denote the set of linear and continuous mappings from X to Y.

Remark 3.37. (a) If X, Y are normed, also $\mathcal{L}(X, Y)$ is normed. The associated 'operator norm' of $T \in \mathcal{L}(X, Y)$ equals[7]

$$\|T\| := \|T\|_{Y \leftarrow X} := \sup_{x \in X \setminus \{0\}} \frac{\|Tx\|_Y}{\|x\|_X}. \tag{3.23}$$

(b) By definition, a continuous map $T : X \to Y$ from $\mathcal{L}(X, Y)$ leads always to a finite supremum (3.23). Vice versa, if a linear operator $T : X \to Y$ yields a finite value in (3.23) (i.e., T is bounded), then T is also continuous.

An example of a Banach space is the set $X = C(D)$ of continuous functions defined on D with bounded norm $\|f\|_\infty := \sup_{x \in D} |f(x)|$ (for a compact D, the supremum is even a maximum).

3.4.7.2 Theorem

The *uniform boundedness theorem* of Banach and Steinhaus connects point-wise and normwise boundedness.

Theorem 3.38. *Assume that*
   (a) X is a Banach space,
   (b) Y is a normed space,
   (c) $\mathcal{T} \subset \mathcal{L}(X, Y)$ is a (in general, infinite) subset of mappings,
   (d) $\sup_{T \in \mathcal{T}} \|Tx\|_Y < \infty$ holds for all $x \in X$.
Then $\mathcal{T}$ is uniformly bounded; i.e., $\sup_{T \in \mathcal{T}} \|T\|_{Y \leftarrow X} < \infty$.

First we add some remarks. Let K be the unit sphere $K := \{x \in X : \|x\| \leq 1\}$. Definition (3.23) states that

$$\|T\|_{Y \leftarrow X} = \sup_{x \in K} \|Tx\|_Y .$$

The statement of the theorem becomes $\sup_{T \in \mathcal{T}} \sup_{x \in K} \|Tx\| < \infty$. Since suprema commute, one may also write $\sup_{x \in K} \sup_{T \in \mathcal{T}} \|Tx\| < \infty$. Assumption (d) of the

---

[7] For the trivial case of $X = \{0\}$, we define the supremum over the empty set by zero.

theorem reads $C(x) := \sup_{T \in \mathcal{T}} \|Tx\| < \infty$; i.e., the function $C(x)$ is point-wise bounded. The astonishing[8] property is that $C(x)$ is even *uniformly* bounded on K.

In the later applications we often apply a particular variant of the theorem.

Corollary 3.39. *Let X and Y be as in Theorem 3.38. Furthermore, assume that the operators $T, T_n \in \mathcal{L}(X, Y)$ $(n \in \mathbb{N})$ satisfy either*

(a) $\{T_n x\}$ a Cauchy sequence for all $x \in X$, or

(b) there exists an operator $T \in \mathcal{L}(X, Y)$ with $T_n x \to Tx$ for all $x \in X$.

Then $\sup_{n \in \mathbb{N}} \|T_n\| < \infty$ holds.

Proof. In this case $\mathcal{T} = \{T_n \in \mathcal{L}(X, Y) : n \in \mathbb{N}\}$ is countably infinite.

Since any Cauchy sequence is bounded, the boundedness $\sup_{n \in \mathbb{N}} \|T_n x\|_Y < \infty$ follows so that Theorem 3.38 is applicable. This proves part (a).

Assumption (b) implies (a). □

3.4.7.3 Proof

The proof of Theorem 3.38 is based on two additional theorems. The first is called *Baire's category theorem* or the *Baire–Hausdorff theorem*.

Theorem 3.40. *Let $X \neq \emptyset$ be a complete metric space. Assume that X has a representation*

$$X = \bigcup_{k \in \mathbb{N}} A_k \qquad \text{with closed sets } A_k.$$

Then there exists at least one $k_0 \in \mathbb{N}$, so that $\mathring{A}_{k_0} \neq \emptyset$ ($\mathring{A}_{k_0}$ denotes the interior of A_{k_0}).

Proof. (a) For an indirect proof assume $\mathring{A}_k = \emptyset$ for all k. We choose a non-empty, open set $U \subset X$ and some $k \in \mathbb{N}$. Since A_k closed, $U \backslash A_k$ is again open and non-empty (otherwise, A_k would contain the open set U; i.e., $\mathring{A}_k \supset U \neq \emptyset$). Since $U \backslash A_k$ is open, it contains a closed sphere $\overline{K_\varepsilon(x)}$ with radius $\varepsilon > 0$ and midpoint x. Without loss of generality, $\varepsilon \leq 1/k$ can be chosen.

(b) Starting with $\varepsilon_0 := 1$ and $x_0 := 0$, according to (a), we choose by induction

$$\overline{K_{\varepsilon_k}(x_k)} \subset K_{\varepsilon_{k-1}}(x_{k-1}) \backslash A_k \quad \text{and} \quad \varepsilon_k \leq 1/k.$$

Since $x_\ell \in K_{\varepsilon_k}(x_k)$ for $\ell \geq k$ and $\varepsilon_k \to 0$ $(k \to \infty)$, $\{x_k\}$ is a Cauchy sequence. Because of completeness, it must converge to some $x := \lim x_k \in X$ and belong to $\overline{K_{\varepsilon_k}(x_k)}$ for all k. Since $\overline{K_{\varepsilon_k}(x_k)} \cap A_k = \emptyset$ by construction, it follows that $x \notin \bigcup_{k \in \mathbb{N}} A_k = X$, which is a contradiction. □

---

[8] Only in the case of a finite-dimensional vector space X, is there a simple proof using $\sup_{T \in \mathcal{T}} \|Tb_i\|_Y < \infty$ for all basis vectors b_i of X.

Theorem 3.41. *Let X be a complete metric space and Y a normed space. For some subset $\mathcal{F} \subset C^0(X,Y)$ of the continuous mappings, assume that $\sup_{f \in \mathcal{F}} \|f(x)\|_Y < \infty$ for all $x \in X$. Then there exist $x_0 \in X$ and $\varepsilon_0 > 0$ such that*

$$\sup_{x \in \overline{K_{\varepsilon_0}(x_0)}} \sup_{f \in \mathcal{F}} \|f(x)\|_Y < \infty. \tag{3.24}$$

Proof. Set $A_k := \bigcap_{f \in \mathcal{F}} \{x \in X : \|f(x)\|_Y \le k\}$ for $k \in \mathbb{N}$ and check that A_k is closed. According to the assumption, each $x \in X$ must belong to some A_k; i.e., $X = \bigcup_{k \in \mathbb{N}} A_k$. Hence the assumptions of Theorem 3.40 are satisfied. Correspondingly, $\overset{\circ}{A}_{k_0} \ne \emptyset$ holds for at least one $k_0 \in \mathbb{N}$. By the definition of A_k, we have $\sup_{x \in A_{k_0}} \sup_{f \in \mathcal{F}} \|f(x)\|_Y \le k_0$. Choose a sphere with $\overline{K_{\varepsilon_0}(x_0)} \subset A_{k_0}$. This yields the desired inequality (3.24) with the bound $\le k_0$. $\square$

For the proof of Theorem 3.38 note that a Banach space is also a complete metric space and $\mathcal{L}(X,Y) \subset C^0(X,Y)$, so that we may set $\mathcal{F} := \mathcal{T}$. The assumption $\sup_{f \in \mathcal{F}} \|f(x)\|_Y < \infty$ is equivalent to $\sup_{T \in \mathcal{T}} \|Tx\|_Y < \infty$. The result (3.24) becomes $\sup_{x \in \overline{K_{\varepsilon_0}(x_0)}} \sup_{T \in \mathcal{T}} \|Tx\|_Y < \infty$. For an arbitrary $\xi \in X \backslash \{0\}$, the element $x_\xi := x_0 + \frac{\varepsilon_0}{\|\xi\|_X} \xi$ belongs to $\overline{K_{\varepsilon_0}(x_0)}$, so that

$$\frac{\|T\xi\|_Y}{\|\xi\|_X} = \frac{1}{\varepsilon_0} \|T(x_\xi - x_0)\|_Y \le \frac{1}{\varepsilon_0} \left(\|Tx_\xi\|_Y + \|Tx_0\|_Y \right)$$

is uniformly bounded for all $T \in \mathcal{T}$ and all $\xi \in X \backslash \{0\}$. Hence the assertion of Theorem 3.38 follows: $\sup_{T \in \mathcal{T}} \sup_{\xi \in X \backslash \{0\}} \frac{\|T\xi\|_Y}{\|\xi\|_X} = \sup_{T \in \mathcal{T}} \|T\| < \infty$.

3.4.8 Necessity of the Stability Condition, Equivalence Theorem

$X = C([0,1])$ together with the maximum norm $\|\cdot\|_\infty$ is a Banach space, and $Y := \mathbb{R}$ is normed (its norm is the absolute value). The mappings

$$f \in C([0,1]) \mapsto I(f) := \int_0^1 f(x)\mathrm{d}x \in \mathbb{R} \quad \text{and} \quad f \in C([0,1]) \mapsto Q_n(f) \in \mathbb{R}$$

are linear and continuous; hence they belong to $\mathcal{L}(X,Y)$. By Remark 3.37b, continuity is equivalent to boundedness, which is quantified in the following lemma.

Lemma 3.42. *The operator norms of I and $Q_n \in \mathcal{L}(X,Y)$ are*

$$\|I\| = 1, \qquad \|Q_n\| = C_n := \sum_{i=0}^n |a_{i,n}|. \tag{3.25}$$

Proof. The estimates $\|I\| \le 1$ and $\|Q_n\| \le C_n$ are equivalent to the estimates (3.12) and (3.13a) from Remark 3.15. According to Remark 3.17b, C_n is the minimal constant, implying $\|Q_n\| = C_n$. The example $f = 1$ shows that one is the best bound for $\|If\|_\infty / \|f\|_\infty$; i.e., $\|I\| = 1$. □

Apply Corollary 3.39 to $T := I$ and $T_n := Q_n$. Convergence of the quadrature formulae $\{Q_n\}$ is expressed by $Q_n(f) \to I(f)$. From Corollary 3.39 we conclude that $\sup_{n \in \mathbb{N}} \|T_n\| < \infty$. According to Lemma 3.42, this means $\sup_{n \in \mathbb{N}} C_n < \infty$ and it is identical to the stability condition of Definition 3.16. Hence, the following theorem is proved.

Theorem 3.43 (stability theorem). *If the family $\{Q_n\}$ of quadrature formulae is convergent, then $\{Q_n\}$ is stable.*

We have already proved 'consistency + stability $\Longrightarrow$ convergence' (cf. (3.22)). Theorem 3.43 yields 'stability $\Longleftarrow$ convergence'. Together, we obtain the following equivalence theorem.

Theorem 3.44 (equivalence theorem). *Assume consistency of the family $\{Q_n\}$ of quadrature formulae. Then stability and convergence are equivalent.*

3.4.9 Modified Definitions for Consistency and Convergence

The terms 'consistency' and 'convergence' can be even better separated, without weakening the previous statements.

The previous definition of convergence contains not only the statement that the sequence $Q_n(f)$ is convergent, but also that it has the desired integral $\int_0^1 f(x)\mathrm{d}x$ as the limit. The latter part can be omitted:

$$\{Q_n\} \text{ is } convergent, \text{ if } \lim_{n \to \infty} Q_n(f) \text{ exists for all } f \in C([0,1]). \qquad (3.26)$$

So far, the definition of consistency is connected with polynomials. Polynomials come into play since we started from interpolatory quadrature based on polynomial interpolation. An interpolatory quadrature, e.g., based on trigonometric interpolation, would not be consistent in the sense of Definition 3.5. According to the sentence following Theorem 3.28, the decisive property of polynomials is that they are dense in $C([0,1])$. One may replace the polynomials by any other dense subset. This leads us to the following generalisation of the term 'consistency':

$$\{Q_n\} \text{ is } consistent \text{ if there is a dense subset } X_0 \subset C([0,1]) \text{ such that} \qquad (3.27)$$

$$Q_n(f) \to \int_0^1 f(x)\mathrm{d}x \qquad \text{for all } f \in X_0.$$

Note that, simultaneously, we have replaced the exactness $Q_n(f) = \int_0^1 f(x)\mathrm{d}x$ for $n \geq n_0$ by the more general convergence definition (3.26). The stability property remains unchanged.

Then the previous theorem can be reformulated as follows.

Theorem 3.45. *(a) Let $\{Q_n\}$ be consistent in the more general sense of (3.27) and stable. Then $\{Q_n\}$ is convergent in the sense of (3.26) and, furthermore, $\lim_{n\to\infty} Q_n(f) = \int_0^1 f(x)\mathrm{d}x$ is the desired value of the integral.*
(b) Let $\{Q_n\}$ be convergent in the sense of (3.26). Then $\{Q_n\}$ is also stable.
(c) Under the assumption of consistency in the sense of (3.27), stability and convergence (3.26) are equivalent.

Proof. (i) It suffices to show that $\lim_{n\to\infty} Q_n(f) = \int_0^1 f(x)\mathrm{d}x$, since this implies (3.26). Let $f \in C([0,1])$ and $\varepsilon > 0$ be given. Because X_0 from (3.27) is dense, there is a $g \in X_0$ with

$$\|f - g\|_\infty \le \frac{\varepsilon}{2\,(1 + C_{\mathrm{stab}})} \qquad (C_{\mathrm{stab}}\text{: stability constant}).$$

According to (3.27), there is an n_0 such that $\left| Q_n(g) - \int_0^1 g(x)\mathrm{d}x \right| \le \frac{\varepsilon}{2}$ for all $n \ge n_0$. The triangle inequality yields the desired estimate

$$\left| Q_n(f) - \int_0^1 f(x)\mathrm{d}x \right|$$
$$\le |Q_n(f) - Q_n(g)| + \left| Q_n(g) - \int_0^1 g(x)\mathrm{d}x \right| + \left| \int_0^1 g(x)\mathrm{d}x - \int_0^1 f(x)\mathrm{d}x \right|$$
$$\le \underbrace{C_{\mathrm{stab}} \|f - g\|_\infty}_{\|f-g\|_\infty \le \varepsilon/[2(1+C_{\mathrm{stab}})]} + \frac{\varepsilon}{2} + \|f - g\|_\infty \overset{\le}{} \varepsilon.$$

(ii) Convergence in the sense of (3.26) guarantees that $\{Q_n(f)\}$ has a limit for all $f \in C([0,1])$. Alternative (a) of Corollary 3.39 applies and yields

$$\sup_{n\in\mathbb{N}_0} \|Q_n\| = \sup_{n\in\mathbb{N}_0} C_n < \infty$$

proving stability.

(iii) Part (c) follows from Parts (a) and (b). □

Finally, we give a possible application of generalised consistency. To avoid the difficulties arising from the instability of the Newton–Cotes formulae, one often uses compound Newton–Cotes formulae. The best known example is the *compound trapezoidal rule*, which uses the trapezoidal rule on each subinterval $[i/n, (i+1)/n]$. The compound trapezoidal rule defines again a family $\{Q_n\}$. It is not consistent in the sense of Definition 3.5, since except for constant and linear functions, no further polynomials are integrated exactly. Instead, we return to the formulation (3.8) of the quadrature error. The well-known estimate states that

$$\left| Q_n(f) - \int_0^1 f(x)\mathrm{d}x \right| \le \frac{1}{12n^2} \|f''\|_\infty \to 0 \tag{3.28}$$

for all $f \in C^2([0,1])$ (cf. [9, §3.1]). The subset $C^2([0,1])$ is dense in $C([0,1])$ (simplest proof: $C^2([0,1]) \supset \{\text{polynomials}\}$ and the latter set is already dense according to Theorem 3.28). Hence the compound trapezoidal rule $\{Q_n\}$ satisfies the consistency condition (3.27) with $X_0 = C^2([0,1])$. The stability of $\{Q_n\}$ follows from Conclusion 3.19a, since all weights are positive and $Q_n(1) = 1$. From Theorem 3.45a we conclude that $Q_n(f) \to \int_0^1 f(x)\mathrm{d}x$ for all continuous f.

The trapezoidal rule is the Newton–Cotes method for $n = 1$. We may fix any $n_{\mathrm{NC}} \in \mathbb{N}$ and use the corresponding Newton–Cotes formula in each subinterval $[i/n, (i+1)/n]$. Again, this compound formula is stable, where the stability constant is given by $C_{n_{\mathrm{NC}}}$ from (3.25).

3.5 Further Remarks

3.5.1 General Intervals and Product Quadrature

The restriction of the integral to $[0,1]$ is a kind of normalisation. If quadrature is needed over an interval $[a,b]$ of length $L = b - a$, use the affine mapping

$$\phi : [0,1] \to [a,b] \qquad \text{defined by} \quad \phi(x) = a + xL.$$

For $g \in C([a,b])$, we use $\int_a^b g(t)\mathrm{d}t = \int_0^1 f(x)\mathrm{d}x$ with $f(x) := Lg(\phi(x))$ and apply the quadrature Q_n from (3.1) to f:

$$\int_a^b g(t)\mathrm{d}t \approx Q_n\left(Lg(\phi(\cdot))\right) = L \sum_{i=0}^n a_{i,n} g(\phi(x_{i,n})).$$

Obviously, expressed in g evaluations, we obtain a new quadrature on $[a,b]$ by

$$\int_a^b g(t)\mathrm{d}t \approx Q_n^{[a,b]}(g) := L \sum_{i=0}^n a_{i,n} g(t_{i,n}) \text{ with } t_{i,n} := \phi(x_{i,n}) = a + x_{i,n}L.$$

Also the error estimate can be transferred from $[0,1]$ to a general interval $[a,b]$. Assume an error estimate (3.8) for $f \in C^{k_n}([0,1])$ by

$$\left| \int_0^1 f(x)\mathrm{d}x - Q_n(f) \right| \le c_n \|f^{(k_n)}\|_\infty.$$

The transformation from above shows immediately that

$$\left| \int_a^b g(t)\mathrm{d}t - Q_n^{[a,b]}(g) \right| \le c_n L^{k_n+1} \|g^{(k_n)}\|_\infty.$$

The stability constant C_n is the minimal c_n for $k_n = 0$. One sees that C_n in $[0,1]$ becomes $C_n^{[a,b]} := LC_n$ in $[a,b]$. This fact can be interpreted in the way that the relative quadrature error $\frac{1}{L}\left| \int_a^b g(t)\mathrm{d}t - Q_n^{[a,b]}(g)\right|$ possesses an unchanged stability constant. Anyway, the stability properties of $\{Q_n\}$ and $\{Q_n^{[a,b]}\}$ are the same.

In applications it happens that the integrand is a product $f(x)g(x)$, where one factor—say g—is not well-suited for quadrature (it may be insufficiently smooth, e.g., containing a weak singularity or it may be highly oscillatory). Interpolation of f by $I_n(f) = \sum_{i=0}^n f(x_{i,n})\Phi_{i,n}(x)$ (cf. §3.1.2) induces a quadrature of fg by

$$\int_0^1 f(x)g(x)\mathrm{d}x \approx \sum_{i=0}^n a_{i,n}f(x_{i,n}) \qquad \text{with } a_{i,n} := \int_0^1 \Phi_{i,n}(x)g(x)\mathrm{d}x,$$

which requires that we have precomputed the (exact) integrals $\int_0^1 \Phi_{i,n}(x)g(x)\mathrm{d}x$.

3.5.2 Consistency Versus Stability

As we shall see consistency is often restricted by stability requirements (cf. Remark 4.15, §5.5.6, §6.6). In this respect, quadrature is an exceptional case. Gauss quadrature Q_n is optimal with respect to stability (its stability constant C_{stab} has the smallest possible value 1) and it possesses the largest possible consistency order $2n + 1$.

Another astonishing observation is that the Gauss quadrature is stable, although it is an interpolatory quadrature based on the *unstable* polynomial interpolation (cf. §4.5).

3.5.3 Perturbations

In this subsection we consider perturbations of f as well as of Q_n. Instead of a fixed function f, consider a sequence $f_n \to f$ in $C([0,1])$ with the intention of replacing $Q_n(f)$ by $Q_n(f_n)$.

A possible application of this setting may be as follows. Let $t_n \to \infty$ be a sequence of natural numbers and f_n the computer realisation of f by a floating-point arithmetic with mantissa length t_n, so that $\|f_n - f\|_\infty \leq C2^{-t_n}$. Then $Q_n(f_n)$ means that parallel to the increase of the number of quadrature points the arithmetical precision also improves.

The following theorem states that such a perturbation does not destroy convergence.

Theorem 3.46. *Let the family $\{Q_n : n \in \mathbb{N}_0\}$ of quadrature formulae be consistent and stable. Furthermore, assume $f_n \to f$ for a sequence of $f_n \in C([0,1])$. Then also $Q_n(f_n)$ converges to $\int_0^1 f(x)\mathrm{d}x$.*

Proof. The previous considerations guarantee $Q_n(f) \to \int_0^1 f(x)\mathrm{d}x$. Thanks to stability, the perturbation $|Q_n(f_n) - Q_n(f)|$ is bounded by $C_{\text{stab}} \|f_n - f\|_\infty \to 0$ and a zero sequence.   □

Similarly, we may perturb Q_n. For instance, assume that the exact weights $a_{i,n}$ are replaced by the truncation $\widetilde{a_{i,n}}$ to mantissa length t_n. Then $\widetilde{Q_n} = Q_n + \delta Q_n$ holds, where $|\delta Q_n(f)| \le C 2^{-t_n} \|f\|_\infty$. For $t_n \to \infty$ we obtain the norm convergence $\|\delta Q_n\| \to 0$, where

$$\|\delta Q_n\| := \sup\{|\delta Q_n(f)| : f \in C([0,1]) \text{ with } \|f\|_\infty = 1\}.$$

Theorem 3.47. *Let the family* $\{Q_n : n \in \mathbb{N}_0\}$ *of quadrature formulae be consistent and stable. Furthermore, assume* $\|\delta Q_n\| \to 0$ *for a sequence of perturbations* δQ_n. *Then also* $(Q_n + \delta Q_n)(f)$ *converges to* $\int_0^1 f(x)\mathrm{d}x$.

Proof. By definition $\delta Q_n(f) \to 0$, while $Q_n(f) \to \int_0^1 f(x)\mathrm{d}x$.   □

Combining both theorems we even get $(Q_n + \delta Q_n)(f_n) \to \int_0^1 f(x)\mathrm{d}x$.

3.5.4 Arbitrary Slow Convergence Versus Quantitative Convergence

While estimates as in (3.28) describe the convergence in a quantitative form, the previous statement $Q_n(f) \to \int_0^1 f(x)\mathrm{d}x$ says nothing about the speed of convergence. Such non-quantitative convergence statements are not very helpful in numerical applications. If one does not know whether an error $\varepsilon = 0.01$ has already been obtained for $n = 5$, or for $n = 10^6$, or even only for $n = 10^{10^{10}}$, one cannot rely on such numerical methods.

Exercise 3.48. *Given a family of quadrature formulae and any number N, construct a continuous function f with the properties*

$$\|f\|_\infty = 1, \qquad \int_0^1 f(x)\mathrm{d}x \ge 1/2, \qquad but\ Q_n(f) = 0 \quad for\ all\ n \le N.$$

Can something be stated about the convergence speed of $Q_n(f) \to \int_0^1 f(x)\mathrm{d}x$ for a general $f \in C([0,1])$? A quantified version of convergence can be described in two equivalent ways. Either we prescribe an $\varepsilon > 0$ and ask for an $n(\varepsilon)$ such that $|Q_m(f) - \int_0^1 f(x)\mathrm{d}x| \le \varepsilon \|f\|_\infty$ for $m \ge n(\varepsilon)$. Or there is a monotone zero sequence ε_n such that $|Q_n(f) - \int_0^1 f(x)\mathrm{d}x| \le \varepsilon_n \|f\|_\infty$.

Remark 3.9 yields a negative result: if $|Q_n(f) - \int_0^1 f(x)\mathrm{d}x| \le \varepsilon_n \|f\|_\infty$ holds for all $f \in C([0,1])$, necessarily $\varepsilon_n \ge 1$ must hold excluding any zero sequence ε_n. Consequently, the convergence $Q_n(f) \to \int_0^1 f(x)\mathrm{d}x$ can be *arbitrarily slow*.

The quantitative convergence result of inequality (3.28) holds for $f \in C^2([0,1])$; i.e., for smoother functions. In fact, we get a similar result for $|Q_n(f) - \int_0^1 f(x)\mathrm{d}x|$, if we consider, say $f \in C^1([0,1])$, instead of $f \in C([0,1])$. Note that $f \in C^1([0,1])$ comes with the norm $\|f\|_{C^1([0,1])} = \max\{|f(x)|, |f'(x)| : 0 \leq x \leq 1\}$.

The next result is prepared by the following lemma. We remark that a subset B of a Banach space is *precompact* if and only if the closure $\bar{B}$ is compact, which means that all sequences $\{f_k\} \subset B$ possess a convergent subsequence: $\lim_{n \to \infty} f_{k_n} \in \bar{B}$. The term 'precompact' is synonymous with 'relatively compact'.

Lemma 3.49. *Let $M \subset X$ be a precompact subset of the Banach space X. Let the operators $A_n \in \mathcal{L}(X,Y)$ be point-wise convergent to $A \in \mathcal{L}(X,Y)$ (i.e., $A_n x \to A x$ for all $x \in X$). Then the sequences $\{A_n x\}$ converge uniformly for all $x \in M$; i.e.,*

$$\sup_{x \in M} \|A_n x - A x\|_Y \to 0 \qquad for\ n \to \infty. \tag{3.29}$$

Proof. (i) $C := \sup\{\|A_n\| : n \in \mathbb{N}\}$ is finite ('stability', cf. Corollary 3.39). Furthermore, $\|A\| \leq C$ is a simple conclusion.

(ii) We disprove the negation of (3.29). Assume that there are an $\varepsilon > 0$ and a subsequence $\mathbb{N}' \subset \mathbb{N}$ such that $\sup_{x \in M} \|A_n x - A x\|_Z \geq \varepsilon$ for all $n \in \mathbb{N}'$. Therefore, some $x_n \in M$ exists with

$$\|A_n x_n - A x_n\|_Z \geq \varepsilon/2 \qquad for\ all\ n \in \mathbb{N}'.$$

Since M is precompact, there is a further subsequence $\mathbb{N}'' \subset \mathbb{N}'$, so that the limit $\lim_{n \in \mathbb{N}''} x_n =: \xi \in \bar{M}$ exists. Choose $n \in \mathbb{N}''$ with $\|x_n - \xi\|_Y \leq \varepsilon/(8C)$ and $\|A_n \xi - A \xi\|_Z < \varepsilon/4$. For this n we obtain

$$\|A_n x_n - A x_n\|_Z \leq \|(A_n - A)(x_n - \xi)\|_Z + \|(A_n - A)\xi\|$$
$$< \underbrace{(\|A_n\| + \|A\|)}_{\leq 2C} \|y_n - \xi\|_Y + \varepsilon/4 \leq \varepsilon/2$$

in contradiction to the previous inequality. □

Theorem 3.50. *There is a zero sequence $\varepsilon_n \to 0$ such that*[9]

$$\left| Q_n(f) - \int_0^1 f(x)\mathrm{d}x \right| \leq \varepsilon_n \|f\|_{C^1([0,1])} \qquad for\ all\ f \in C^1([0,1]). \tag{3.30}$$

Proof. Let $X = C([0,1])$. The subset $M := \{f \in C^1([0,1]) : \|f\|_{C^1([0,1])}\} \subset X$ is precompact due to the theorem of Arzelà–Ascoli recalled below. Apply Lemma 3.49 with $Y = \mathbb{R}$, $A_n = Q_n$, and $A(f) = \int_0^1 f(x)\mathrm{d}x$ (note that $\mathcal{L}(X,\mathbb{R}) = X^*$). Set

$$\varepsilon_n := \sup_{f \in M} \left| Q_n(f) - \int_0^1 f(x)\mathrm{d}x \right|$$

A simple scaling argument shows that (3.30) holds. Furthermore, Lemma 3.49 states that $\varepsilon_n \to 0$. □

---

[9] We may replace $C^1([0,1])$ in Theorem 3.50 by the Hölder space $C^\delta([0,1])$ for any exponent $\delta > 0$. Its norm is $\|f\|_{C^\delta([0,1])} = \max\{\|f\|_\infty, \sup_{x \neq y} |f(x) - f(y)|/|x - y|^\delta\}$.

Theorem 3.51 (Arzelà–Ascoli). *Let D be compact and $M \subset C(D)$. Suppose that M is uniformly bounded:*

$$\sup\{\|f\|_{C(D)} : f \in M\},$$

and equicontinuous; i.e., for any $\varepsilon > 0$ and $x \in D$, there is some δ such that

$$|f(x) - f(y)| \le \varepsilon \qquad \text{for all } f \in M, \text{ and all } x, y \in D \text{ with } |x - y| \le \delta.$$

Then M is precompact.

For a proof see Yosida [12, 1,§III.3].

References

1. Bulirsch, R.: Bemerkungen zur Romberg-Integration. Numer. Math. **6**, 6–16 (1964)
2. Christoffel, E.B.: Über die Gaußische Quadratur und eine Verallgemeinerung derselben. J. Reine Angew. Math. **55**, 61–82 (1858)
3. Davis, P.J., Rabinowitz, P.: Methods of Numerical Integration, 2nd ed. Academic Press, New York (1984)
4. Gauss, C.F.: Methodus nova integralium valores per approximationem inveniendi. In: Werke, vol. 3, pp. 163–196. K. Gesellschaft Wissenschaft, Göttingen (1876). (reprint by Georg Olms, Hildesheim, 1981)
5. Natanson, I.P.: Konstruktive Funktionentheorie. Akademie-Verlag, Berlin (1955)
6. Ouspensky, J.V.: Sur les valeurs asymptotiques des coefficients de Cotes. Bull. Amer. Math. Soc. **31**, 145–156 (1925)
7. Quarteroni, A., Sacco, R., Saleri, F.: Numerical Mathematics, 2nd ed. Springer, Berlin (2007)
8. Romberg, W.: Vereinfachte numerische Quadratur. Norske Vid. Selsk. Forh. [Proceedings of the Royal Norwegian Society of Sciences and Letters], Trondheim **28**, 30–36 (1955)
9. Stoer, J., Bulirsch, R.: Introduction to Numerical Analysis. North-Holland, Amsterdam (1980)
10. Stroud, A., Secrest, D.: Gaussian Quadrature Formulas. Prentice-Hall, Englewood Cliffs (1966)
11. Whiteside, D.T. (ed.): The mathematical papers of Isaac Newton, Vol. 4. Cambridge University Press, Cambridge (1971)
12. Yosida, K.: Functional Analysis. Springer, New York (1968)
13. Zeidler, E. (ed.): Oxford Users' Guide to Mathematics. Oxford University Press, Oxford (2004)

Chapter 4
Interpolation

Interpolation by polynomials is a field in which stability issues have been addressed quite early. Section 4.5 will list a number of classical results.

4.1 Interpolation Problem

The usual *linear*[1] *interpolation problem* is characterised by a subspace V_n of the Banach space $C([0, 1])$ (with norm $\|\cdot\|_\infty$; cf. §3.4.7.1) and a set

$$\{x_{i,n} \in [0, 1] : 0 \le i \le n\}$$

of $n + 1$ different[2] interpolation points, also called nodal points. Given a tuple $\{y_i : 0 \le i \le n\}$ of 'function values', an interpolant $\Phi \in V_n$ with the property

$$\Phi(x_{i,n}) = y_i \qquad (0 \le i \le n) \tag{4.1}$$

has to be determined.

Exercise 4.1. *(a) The interpolation problem is solvable for all tuple $\{y_i : 0 \le i \le n\}$, if and only if the linear space*

$$\mathcal{V}_n := \left\{ (\Phi(x_{i,n}))_{i=0}^n \in \mathbb{R}^{n+1} : \Phi \in V_n \right\}$$

has dimension $n + 1$.
(b) If $\dim \mathcal{V}_n = n + 1$, the interpolation problem is uniquely *solvable.*

The interpolation problem (4.1) can be reduced to a system of $n + 1$ linear equations. As is well known, there are two alternatives for linear systems:

---

[1] The term 'linear' refers to the underlying *linear* space V_n, not to linear functions.
[2] In the case of the more general Hermite interpolation, a p-fold interpolation point ξ corresponds to prescribed values of the derivatives $f^{(m)}(\xi)$ for $0 \le m \le p - 1$.

W. Hackbusch, *The Concept of Stability in Numerical Mathematics*,
Springer Series in Computational Mathematics 45, DOI 10.1007/978-3-642-39386-0_4,
© Springer-Verlag Berlin Heidelberg 2014

(a) either the interpolation problem is uniquely solvable for *arbitrary* values y_i or
(b) the interpolant either does not exist for certain y_i or is not unique.

The *polynomial interpolation* is characterised by

$$V_n = \{\text{polynomials of degree } \leq n\}$$

and is always solvable. In the case of general vector spaces V_n, we always assume
that the interpolation problem is uniquely solvable.

For the special values $y_i = \delta_{ij}$ (j fixed, δ_{ij} Kronecker symbol), one obtains an
interpolant $\Phi_{j,n} \in V_n$, which we call the j-th *Lagrange function* (analogous to the
Lagrange polynomials in the special case of polynomial interpolation).

Exercise 4.2. *(a) The interpolant for arbitrary y_i $(0 \leq i \leq n)$ is given by*

$$\Phi = \sum_{i=0}^{n} y_i \Phi_{i,n} \in V_n. \tag{4.2}$$

(b) In the case of polynomial interpolation, the Lagrange polynomial is defined by

$$L_{i,n}(x) := \Phi_{i,n}(x) := \prod_{j \in \{0,\dots n\}\backslash\{i\}} \frac{x - x_j}{x_i - x_j}. \tag{4.3}$$

For continuous functions f we define

$$I_n(f) := \sum_{i=0}^{n} f(x_{i,n}) \Phi_{i,n} \tag{4.4}$$

as interpolant of $y_i = f(x_{i,n})$. Hence

$$I_n : C([0,1]) \to C([0,1])$$

is a linear mapping from the continuous functions into itself.

Exercise 4.3. *(a) The interpolation $I_n : X = C([0,1]) \to C([0,1])$ is continuous
and linear; i.e., $I_n \in \mathcal{L}(X,X)$.*
(b) I_n is a projection; i.e., $I_n I_n = I_n$.

The terms 'convergence', 'consistency' and 'stability' of the previous chapter
can easily be adapted to the interpolation problem. Note that we have not only one
interpolation I_n, but a family $\{I_n : n \in \mathbb{N}_0\}$ of interpolations.

The interval $[0,1]$ is chosen without loss of generality. The following results
can immediately be transferred to a general interval $[a,b]$ by means of the affine
mapping $\phi(t) = (t - a)/(b - a)$. The Lagrange functions $\Phi_{i,n} \in C([0,1])$ become
$\hat{\Phi}_{i,n} := \Phi_{i,n} \circ \phi \in C([a,b])$. Note that in the case of polynomials, $\Phi_{i,n}$ and $\hat{\Phi}_{i,n}$
have the same polynomial degree n. The norms $\|I_n\|$ and the stability constant C_{stab}
from §4.3 do not change! Also the error estimate (4.8) remains valid.

Another subject are interpolations on higher-dimensional domains $D \subset \mathbb{R}^d$. The general concept is still true, but the concrete one-dimensional interpolation methods do not necessarily have a counterpart in d dimensions. An exception are domains which are Cartesian products. Then one can apply the tensor product interpolation discussed in §4.7.

4.2 Convergence and Consistency

Definition 4.4 (convergence). *A family* $\{I_n : n \in \mathbb{N}_0\}$ *of interpolations is called* convergent *if*
$$\lim_{n \to \infty} I_n(f) \text{ exists for all } f \in C([0,1]).$$

Of course, we intend that $I_n(f) \to f$, but here convergence can be defined without fixing the limit, since $\lim I_n(f) = f$ will come for free due to consistency.

Concerning consistency, we follow the model of (3.27).

Definition 4.5 (consistency). *A family* $\{I_n : n \in \mathbb{N}_0\}$ *of interpolations is called* consistent *if there is a dense subset* $X_0 \subset C([0,1])$ *such that*

$$I_n(g) \to g \qquad \text{for all } g \in X_0.$$

Exercise 4.6. *Let* $\{I_n\}$ *be the interpolation by polynomials of degree* $\leq n$. *Show that a possible choice of the dense set in Definition 4.5 is* $X_0 := \{polynomials\}$.

4.3 Stability

First, we characterise the operator norm $\|I_n\|$ (cf. (3.23)).

Lemma 4.7. $\|I_n\| = \left\|\sum_{i=0}^n |\Phi_{i,n}(\cdot)|\right\|_\infty$ *holds with* $\Phi_{i,n}$ *from (4.4).*

Proof. (i) Set $C_n := \left\|\sum_{i=0}^n |\Phi_{i,n}(\cdot)|\right\|_\infty$. For arbitrary $f \in C([0,1])$ we conclude that

$$|I_n(f)(x)| = \left|\sum_{i=0}^n f(x_{i,n})\Phi_{i,n}(x)\right| \leq \sum_{i=0}^n \underbrace{|f(x_{i,n})|}_{\leq \|f\|_\infty} |\Phi_{i,n}(x)| \leq \|f\|_\infty \sum_{i=0}^n |\Phi_{i,n}(x)|$$
$$\leq \|f\|_\infty C_n.$$

Since this estimate holds for all $x \in [0,1]$, it follows that $\|I_n(f)\| \leq C_n \|f\|_\infty$. Because f is arbitrary, $\|I_n\| \leq C_n$ is proved.

(ii) Let the function $\sum_{i=0}^n |\Phi_{i,n}(\cdot)|$ be maximal at x_0: $\sum_{i=0}^n |\Phi_{i,n}(x_0)| = C_n$. Choose $f \in C([0,1])$ with $\|f\|_\infty = 1$ and $f(x_{i,n}) = \text{sign}(\Phi_{i,n}(x_0))$. Then

$$|I_n(f)(x_0)| = \left| \sum_{i=0}^{n} f(x_{i,n}) \Phi_{i,n}(x_0) \right| = \sum_{i=0}^{n} |\Phi_{i,n}(x_0)| = C_n = C_n \|f\|_\infty$$

holds; i.e., $\|I_n(f)\|_\infty = C_n \|f\|_\infty$ for this f. Hence the operator norm

$$\|I_n\| = \sup \{ \|I_n(f)\|_\infty / \|f\|_\infty : f \in C([0,1]) \backslash \{0\} \}$$

is bounded from below by $\|I_n\| \geq C_n$. Together with (i), the equality $\|I_n\| = C_n$ is proved. $\square$

Again, stability expresses the boundedness of the sequence of norms $\|I_n\|$.

Definition 4.8 (stability). *A family $\{I_n : n \in \mathbb{N}_0\}$ of interpolations is called* stable *if*

$$C_{\text{stab}} := \sup_{n \in \mathbb{N}_0} \|I_n\| < \infty \qquad \text{for} \quad \|I_n\| = \left\| \sum_{i=0}^{n} |\Phi_{i,n}(\cdot)| \right\|_\infty . \qquad (4.5)$$

In the context of interpolation, the stability constant C_{stab} is called *Lebesgue constant*.

Polynomial interpolation is a particular way to approximate a continuous function by a polynomial. Note that the more general approximation due to Weierstrass is convergent. The relation between the best possible polynomial approximation and the polynomial interpolation is considered next.

Remark 4.9. *Given $f \in C([0,1])$, let p_n^* be the best approximation to f by a polynomial[3] of degree $\leq n$, while p_n is its interpolant. Then the following estimate holds:*

$$\|f - p_n\| \leq (1 + C_n) \|f - p_n^*\| \qquad \text{with } C_n = \|I_n\| . \qquad (4.6)$$

Proof. Any polynomial of degree $\leq n$ is reproduced by interpolation, in particular, $I_n p_n^* = p_n^*$. Hence,

$$f - p_n = f - I_n f = f - [I_n(f - p_n^*) + I_n p_n^*] = f - p_n^* + I_n(f - p_n^*)$$

can be estimated as claimed above. $\square$

Note that by the Weierstrass approximation theorem 3.28,

$$\|f - p_n^*\| \to 0$$

holds. An obvious conclusion from (4.6) is the following: If stability would hold (i.e., $C_n \leq C_{\text{stab}}$), also $\|f - p_n\| \to 0$ follows. Instead, we shall show instability, and the asymptotic behaviour on the right-hand side in (4.6) depends on which process is faster: $\|f - p_n^*\| \to 0$ or $C_n \to \infty$.

---

[3] The space of polynomials can be replaced by any other interpolation subspace V_n.

4.4 Equivalence Theorem

Following the scheme (3.22), we obtain the next statement.

Theorem 4.10 (convergence theorem). *Assume that the family $\{I_n : n \in \mathbb{N}_0\}$ of interpolations is consistent and stable. Then it is also convergent, and furthermore, $I_n(f) \to f$ holds.*

Proof. Let $f \in C([0, 1])$ and $\varepsilon > 0$ be given. There is some $g \in X_0$ with

$$\|f - g\|_\infty \leq \frac{\varepsilon}{2(1 + C_{\text{stab}})},$$

where C_{stab} is the stability constant. According to Definition 4.5, there is an n_0 such that $\|I_n(g) - g\|_\infty \leq \frac{\varepsilon}{2}$ for all $n \geq n_0$. The triangle inequality yields the desired estimate:

$$\|I_n(f) - f\|_\infty \leq \|I_n(f) - I_n(g)\|_\infty + \|I_n(g) - g\|_\infty + \|g - f\|_\infty$$
$$\leq C_{\text{stab}} \underbrace{\|f - g\|_\infty}_{\|f-g\|_\infty \leq \varepsilon/[2(1+C_{\text{stab}})]} + \frac{\varepsilon}{2} + \|f - g\|_\infty \quad \leq \quad \varepsilon. \qquad \square$$

Again, the stability condition turns out to be necessary.

Lemma 4.11. *A convergent family $\{I_n : n \in \mathbb{N}_0\}$ of interpolations is stable.*

Proof. Since $\{I_n(f)\}$ converges, the I_n are uniformly bounded. Apply Corollary 3.39 with $X = Y = C([0, 1])$ and $T_n := I_n \in \mathcal{L}(X; Y)$. $\square$

Theorem 4.10 and Lemma 4.11 yield the following equivalence theorem.

Theorem 4.12. *Let the family $\{I_n : n \in \mathbb{N}_0\}$ of interpolations be consistent. Then convergence and stability are equivalent.*

4.5 Instability of Polynomial Interpolation

We choose the equidistant interpolation points $x_{i,n} = i/n$ and restrict ourselves to even n. The Lagrange polynomial $L_{\frac{n}{2},n}$ is particularly large in the subinterval $(0, 1/n)$. In its midpoint we observe the value

$$\left| L_{\frac{n}{2},n}\left(\tfrac{1}{2n}\right) \right| = \left| \prod_{\substack{j=0 \\ j \neq \frac{n}{2}}}^{n} \frac{\frac{1}{2n} - \frac{j}{n}}{\frac{1}{2} - \frac{j}{n}} \right| = \left| \prod_{\substack{j=0 \\ j \neq \frac{n}{2}}}^{n} \frac{\frac{1}{2} - j}{\frac{n}{2} - j} \right|$$
$$= \frac{\frac{1}{2} \times \frac{1}{2} \times \frac{3}{2} \times \ldots \times \left(\frac{n}{2} - \frac{3}{2}\right) \times \left(\frac{n}{2} + \frac{1}{2}\right) \times \ldots \times \left(n - \frac{1}{2}\right)}{\left[\left(\frac{n}{2}\right)!\right]^2}.$$

Exercise 4.13. *Show that the expression from above diverges exponentially.*

Because of $C_n = \|\sum_{i=0}^n |L_{i,n}(\cdot)|\|_\infty \geq \|L_{\frac{n}{2},n}\|_\infty \geq |L_{\frac{n}{2},n}(\frac{1}{2n})|$, interpolation (at equidistant interpolation points) cannot be stable. The true behaviour of C_n has first[4] been described by Turetskii [21]:

$$C_n \approx \frac{2^{n+1}}{e\,n\log n}.$$

The asymptotic is improved by Schönhage [18, Satz 2] to[5]

$$C_n \approx 2^{n+1}/[e\,n\,(\gamma + \log n)],$$

where γ is Euler's constant.[6] Even more asymptotic terms are determined in [11].

One may ask whether the situation improves for another choice of interpolation points. In fact, an asymptotically optimal choice are the so-called *Chebyshev points*:

$$x_{i,n} = \frac{1}{2}\left(1 + \cos\left(\tfrac{i+1/2}{n+1}\pi\right)\right)$$

(these are the zeros of the Chebyshev polynomial[7] $T_{n+1} \circ \phi$, where $\phi(\xi) = 2\xi + 1$ is the affine transformation from $[0, 1]$ onto $[-1, 1]$). In this case, one can prove that[8]

$$\|I_n\| \leq 1 + \frac{2}{\pi}\log(n+1) \tag{4.7}$$

(cf. Rivlin [17, Theorem 1.2]), which is asymptotically the best bound, as the next result shows.

Theorem 4.14. *There is some $c > 0$ such that*

$$\|I_n\| > \frac{2}{\pi}\log(n+1) - c$$

holds for any choice of interpolation points.

In 1914, Faber [6] proved

$$\|I_n\| > \frac{1}{12}\log(n+1),$$

while, in 1931, Bernstein [1] showed the asymptotic estimate

---

[4] For historical comments see [20].

[5] The function $\varphi = \sum_{i=0}^n |L_{i,n}(\cdot)|$ attains its maximum C_n in the first and last interval. As pointed out by Schönhage [18, §4], φ is of similar size as in (4.7) for the middle interval.

[6] The value $\gamma = 0.5772\ldots$ is already given in Euler's first article [5]. Later, Euler computed 15 exact decimals places of γ.

[7] The Chebyshev polynomial $T_n(x) := \cos(n\arccos(x))$, $n \in \mathbb{N}_0$, satisfies the three-term recursion $T_{n+1}(x) = 2xT_n(x) - T_{n-1}(x)$ $(n \geq 1)$, starting from $T_0(x) = 1$ and $T_1(x) = x$.

[8] A lower bound is $\|I_n\| > \frac{2}{\pi}\log(n+1) + \frac{2}{\pi}(\gamma + \log\frac{8}{\pi}) = \lim_{n\to\infty} \|I_n\|$, where $\frac{2}{\pi}(\gamma + \log\frac{8}{\pi}) = 0.962\,52\ldots$

$$\|I_n\| > \frac{2-\varepsilon}{\pi}\log(n+1) \qquad \text{for all } \varepsilon > 0.$$

The estimate of Theorem 4.14 originates from Erdös [4]. The bound

$$\|I_n\| > \frac{1}{8\sqrt{\pi}}\log(n+1)$$

can be found in Natanson [12, p. 370f].

The idea of the proof is as follows. Given $x_{i,n} \in [0,1]$, $0 \le i \le n$, construct a polynomial P of degree $\le n$ (concrete construction, e.g., in [12, p. 370f], [13]) such that $|P(x_{i,n})| \le 1$, but $P(\xi) > M_n$ for at least one point $\xi \in [0,1]$. Since the interpolation of P is exact, i.e., $I_n(P) = P$, the evaluation at ξ yields

$$\|I_n\| = \left\|\sum_{i=0}^{n} |L_{i,n}(\cdot)|\right\|_\infty \ge \sum_{i=0}^{n} |L_{i,n}(\xi)|$$

$$\ge \sum_{i=0}^{n} |P(x_{i,n})L_{i,n}(\xi)| \ge \left|\sum_{i=0}^{n} P(x_{i,n})L_{i,n}(\xi)\right| = |P(\xi)| > M_n,$$

proving $\|I_n\| > M_n$.

We conclude that any sequence of polynomial interpolations I_n is unstable.

4.6 Is Stability Important for Practical Computations?

Does the instability of polynomial interpolation mean that one should avoid polynomial interpolation altogether? Practically, one may be interested in an interpolation I_{n^*} for a *fixed* n^*. In this case, the theoretically correct answer is: the property of I_{n^*} has nothing to do with convergence and stability of $\{I_n\}_{n\in\mathbb{N}}$. The reason is that convergence and stability are asymptotic properties of the sequence $\{I_n\}_{n\in\mathbb{N}}$ and are in no way related to the properties of a particular member I_{n^*} of the sequence. One can construct two different sequences $\{I'_n\}_{n\in\mathbb{N}}$ and $\{I''_n\}_{n\in\mathbb{N}}$—one stable, the other unstable—such that $I'_{n^*} = I''_{n^*}$ belongs to both sequences. This argument also holds for the quadrature discussed in the previous chapter.

On the other hand, we may expect that instability expressed by $C_n \to \infty$ may lead to large values of C_n, unless n is very small. We return to this aspect later.

The convergence statement from Definition 4.4 is, in practice, of no help. The reason is that the convergence from Definition 4.4 can be arbitrarily slow, so that for a fixed n, it yields no hint concerning the error $I_n(f) - f$. Reasonable error estimates can only be given if f has a certain smoothness, e.g., $f \in C^{n+1}([0,1])$. Then the standard error estimate of polynomial interpolation states that

$$\|f - I_n(f)\|_\infty \le \frac{1}{(n+1)!} C_\omega(I_n) \left\|f^{(n+1)}\right\|_\infty, \tag{4.8}$$

where

$$C_\omega(I_n) := \|\omega\|_\infty \quad \text{for } \omega(x) := \prod_{i=0}^{n} (x - x_{i,n})$$

(cf. [14, §1.5], [19], [15, §8.1.1], [8, §B.3]). The quantity $C_\omega(I_n)$ depends on the location of the interpolation points. It is minimal for the Chebyshev points, where

$$C_\omega(I_n) = 4^{-(n+1)}.$$

In spite of the instability of polynomial interpolation, we conclude from estimate (4.8) that convergence holds, provided that $\|f^{(n+1)}\|_\infty$ does not grow too much as $n \to \infty$ (of course, this requires that f be analytic). However, in this analysis we have overlooked the numerical rounding errors of the input data.[9] When we evaluate the function values $f(x_{i,n})$, a perturbed result $f(x_{i,n}) + \delta_{i,n}$ is returned with an error $|\delta_{i,n}| \le \eta \|f\|_\infty$. Therefore, the true interpolant is $I_n(f) + \delta I_n$ with $\delta I_n = \sum_{i=0}^{n} \delta_{i,n} \Phi_{i,n}$. An estimate of δI_n is given by $\eta \|I_n\| \|f\|_\infty$. This yields the error estimate

$$\|f - I_n(f) - \delta I_n\|_\infty \le \varepsilon_n^{\text{int}} + \varepsilon_n^{\text{per}} \quad \text{with}$$

$$\varepsilon_n^{\text{int}} = \frac{1}{(n+1)!} C_\omega(I_n) \|f^{(n+1)}\|_\infty \quad \text{and} \quad \varepsilon_n^{\text{per}} = \eta \|I_n\| \|f\|_\infty.$$

Since η is small (maybe of the size of machine precision), the contribution $\varepsilon_n^{\text{int}}$ is not seen in the beginning. However, with increasing n, the part $\varepsilon_n^{\text{int}}$ is assumed to tend to zero, while $\varepsilon_n^{\text{per}}$ increases to infinity because of the instability of I_n.

We illustrate this situation in two different scenarios. In both cases we assume that the analytic function f is such that the exact interpolation error (4.8) decays like $\varepsilon_n^{\text{int}} = e^{-n}$.

(1) Assume a perturbation error $\varepsilon_n^{\text{per}} = \eta e^n$ due to an exponential increase of $\|I_n\|$. The resulting error is

$$e^{-n} + \eta e^n.$$

Regarding n as a real variable, we find a minimum at $n = \frac{1}{2} \log \frac{1}{\eta}$ with the value $2\sqrt{\eta}$. Hence, we cannot achieve better accuracy than half the mantissa length.

(2) According to (4.7), we assume that $\varepsilon_n^{\text{int}} = \eta(1 + \frac{2}{\pi} \log(n+1))$, so that the sum

$$e^{-n} + \eta(1 + \frac{2}{\pi} \log(n+1))$$

is the total error. Here, minimising n is the solution to the fixed-point equation $n = \log(n+1) - \log(2\eta/\pi)$. For $\eta = 10^{-16}$ the minimal value 3.4η of the total error is taken at the integer value $n = 41$. The precision corresponds to almost the full mantissa length. Hence, in this case the instability $\|I_n\| \to \infty$ is completely harmless.[10]

---

[9] There are further rounding errors, which we ignore to simplify the analysis.

[10] To construct an example, where even for (4.7) the instability becomes obvious, one has to assume that the interpolation error decreases very slowly like $\varepsilon_n^{\text{int}} = 1/\log(n)$.

4.7 Tensor Product Interpolation

Finally, we give an example where the norm $\|I_n\|$ is required for the analysis of the interpolation error, even if we ignore input errors and rounding errors. Consider the function $f(x, y)$ in two variables $(x, y) \in [0, 1] \times [0, 1]$. The two-dimensional polynomial interpolation can easily be constructed from the previous I_n. The tensor product[11] $I_n^2 := I_n \otimes I_n$ can be applied as follows. First, we apply the interpolation with respect to x. For any $y \in [0, 1]$ we have

$$F(x, y) := I_n(f(\cdot, y))(x) = \sum_{i=0}^{n} f(x_{i,n}, y)\, \Phi_{i,n}(x).$$

In a second step, we apply I_n with respect to y:

$$I_n^2(f)(x, y) = I_n\left(F(x, \cdot)\right)(y) = \sum_{i=0}^{n}\sum_{j=0}^{n} f(x_{i,n}, x_{j,n})\, \Phi_{i,n}(x)\, \Phi_{j,n}(y).$$

Inequality (4.8) yields a first error[12]

$$|f(x, y) - F(x, y)| \le \frac{1}{(n+1)!} C_\omega(I_n) \left\| \frac{\partial^{n+1}}{\partial x^{n+1}} f \right\|_\infty \qquad \text{for all } x, y \in [0, 1].$$

The second one is

$$F(x, y) - I_n^2(f)(x, y) = \sum_{i=0}^{n} |f(x_{i,n}, y) - I_n(f(x_{i,n}, \cdot)(y)|\, \Phi_{i,n}(x).$$

Again

$$|f(x_{i,n}, y) - I_n(f(x_{i,n}, \cdot))(y)| \le \frac{1}{(n+1)!} C_\omega(I_n) \left\| \frac{\partial^{n+1}}{\partial y^{n+1}} f \right\|_\infty$$

holds and leads us to the estimate

$$\left\| F - I_n^2(f) \right\|_\infty \le \|I_n\| \frac{1}{(n+1)!} C_\omega(I_n) \left\| \frac{\partial^{n+1}}{\partial y^{n+1}} f \right\|_\infty.$$

The previous estimates and the triangle inequality yield the final estimate

$$\left\| f - I_n^2(f) \right\|_\infty \le \frac{1}{(n+1)!} C_\omega(I_n) \left[\|I_n\| \left\| \frac{\partial^{n+1}}{\partial y^{n+1}} f \right\|_\infty + \left\| \frac{\partial^{n+1}}{\partial x^{n+1}} f \right\|_\infty \right].$$

Note that the divergence of $\|I_n\|$ can be compensated by $\frac{1}{(n+1)!}$.

---

[11] Concerning the tensor notation see [9].

[12] Here, $\|\cdot\|_\infty$ is the maximum norm over $[0, 1]^2$.

4.8 Stability of Piecewise Polynomial Interpolation

One possibility to obtain stable interpolations is by constructing a piecewise polynomial interpolation. Here, the degree of the piecewise polynomials is fixed, while
the size of the subintervals approaches zero as $n \to \infty$. Let $J = [0, 1]$ be the
underlying interval. The subdivision is defined by $\Delta_n := \{x_0, x_1, \ldots, x_n\} \subset J$
containing points satisfying

$$0 = x_0 < x_1 < \ldots < x_{n-1} < x_n = 1.$$

This defines the subintervals $J_k := [x_{k-1}, x_k]$ of length $h_k := x_k - x_{k-1}$ and
$\delta_n := \max_{1 \le k \le n} h_k$. In principle, all quantities x_k, J_k, h_k should carry an
additional index n, since each subdivision of the sequence $(\Delta_n)_{n \in \mathbb{N}}$ has different
$x_k = x_k^{(n)}$. For the sake of simplicity we omit this index, except for the grid size
δ_n, which has to satisfy $\delta_n \to 0$.

Among the class of piecewise polynomial interpolations, we can distinguish two
types depending on the support[13] of the Lagrange functions $\Phi_{j,n}$. In case of Type I,
$\Phi_{j,n}$ has a local support, whereas $\text{supp}(\Phi_{j,n}) = J$ for Type II. The precise definition
of a local support is: there are $\alpha, \beta \in \mathbb{N}_0$ independent of n such that

$$\text{supp}(\Phi_{j,n}) \subset \bigcup_{k=\max\{1, j-\alpha\}}^{\min\{n, j+\beta\}} J_k. \tag{4.9}$$

4.8.1 Case of Local Support

The simplest example is the linear interpolation where $I_n(f)(x_k) = f(x_k)$ and $f|_{J_k}$
(i.e., f restricted to J_k) is a linear polynomial. The corresponding Lagrange function
$\Phi_{j,n}$ is called the *hat function* and has the support[14] $\text{supp}(\Phi_{j,n}) = J_j \cup J_{j+1}$.

We may fix another polynomial degree d and fix points $0 = \xi_0 < \xi_1 < \ldots <
\xi_d = 1$. In each subinterval $J_k = [x_{k-1}, x_k]$ we define interpolation nodes $\zeta_\ell :=
x_{k-1} + (x_k - x_{k-1}) \xi_\ell$. Interpolating f by a polynomial of degree d at these nodes,
we obtain $I_n(f)|_{J_k}$. Altogether, $I_n(f)$ is a continuous[15] and piecewise polynomial
function on J. Again, $\text{supp}(\Phi_{j,n}) = J_j \cup J_{j+1}$ holds.

A larger but still local support occurs in the following construction of piecewise
cubic functions. Define $I_n(f)|_{J_k}$ by cubic interpolation at the nodes[14] x_{k-2}, x_{k-1},
x_k, x_{k+1}. Then the support $\text{supp}(\Phi_{j,n}) = J_{j-1} \cup J_j \cup J_{j+1} \cup J_{j+2}$ is larger than
before.

---

[13] The support of a function f defined on I is the closed set $\text{supp}(f) := \overline{\{x \in I : f(x) \neq 0\}}$.

[14] The expression has to be modified for the indices 1 and n at the end points.

[15] If $I_n(f) \in C^1(I)$ is desired, one may use Hermite interpolation; i.e., also $d I_n(f)/dx = f'$ at
$x = x_{k-1}$ and $x = x_k$. This requires a degree $d \ge 3$.

The error estimates can be performed for each subinterval separately. Transformation of inequality (4.8) to J_k yields $\|I_n(f) - f\|_{\infty, J_k} \leq Ch_k^{-(d+1)}\|f^{(d+1)}\|_\infty$, where d is the (fixed) degree of the local interpolation polynomial. The overall estimate is

$$\|I_n(f) - f\|_\infty \leq C\delta_n^{-(d+1)}\|f^{(d+1)}\|_\infty \to 0, \tag{4.10}$$

where we use the condition $\delta_n \to 0$.

Stability is controlled by the maximum norm of $\Phi_n := \sum_{i=1}^n |\Phi_{i,n}(\cdot)|$. For the examples from above it is easy to verify that $\|\Phi_{i,n}\| \leq K$ independently of i and n. Fix an argument $x \in I$. The local support property (4.9) implies that $\Phi_{i,n}(x) \neq 0$ holds for at most $\alpha + \beta + 1$ indices i. Hence $\sum_{i=1}^n |\Phi_{i,n}(x)| \leq \overline{C}_{\text{stab}} := (\alpha + \beta + 1)K$ holds and implies $\sup_n \|I_n\| \leq \overline{C}_{\text{stab}}$ (cf. (4.5)).

4.8.2 Spline Interpolation as an Example for Global Support

The space V_n of the *natural cubic splines* is defined by

$$V_n = \left\{ f \in C^2(I) : f''(0) = f''(1) = 0, \ f|_{J_k} \text{ cubic polynomial for } 1 \leq k \leq n \right\}.$$

The interpolating spline function $S \in V_n$ has to satisfy $S(x_k) = f(x_k)$ for $0 \leq k \leq n$. We remark that S is also the minimiser of

$$\min\left\{ \int_J |g''(x)|^2 \, dx : g \in C^2(J) : S(x_k) = f(x_k) \text{ for } 0 \leq k \leq n \right\}.$$

In this case the support of a Lagrange function $\Phi_{j,n}$, which now is called a *cardinal spline*, has global support:[16] $\operatorname{supp}(\Phi_{j,n}) = J$. Interestingly, there is another basis of V_n consisting of so-called *B-splines* B_j, whose support is local:[14] $\operatorname{supp}(B_j) = J_{j-1} \cup J_j \cup J_{j+1} \cup J_{j+2}$. Furthermore, they are non-negative and sum up to

$$\sum_{j=0}^n B_j = 1. \tag{4.11}$$

We choose an equidistant[17] grid; i.e., $J_i = [(i-1)h, ih]$ with $h := 1/n$. The stability estimate $\|I_n\| = \|\sum_{i=0}^n |\Phi_{i,n}(\cdot)|\|_\infty \leq C_{\text{stab}}$ (cf. (4.5)) is equivalent to

$$\|S\|_\infty \leq C_{\text{stab}} \|y\|_\infty, \quad \text{where } S = \sum_{i=0}^n y_i \Phi_{i,n} \in V_n$$

is the spline function interpolating $y_i = S(x_i)$. In the following, we make use of the

---

[16] $\Phi_{j,n}$ is non-negative in $J_j \cup J_{j+1}$ and has oscillating signs in neighbouring intervals. One can prove that the maxima of $\Phi_{j,n}$ in J_k are exponentially decreasing with $|j - k|$. This fact can already be used for a stability proof.

[17] For the general case compare [14, §2], [15, §8.7], [19, §2.4].

B-splines, which easily can be described for the equidistant case.[18] The evaluation at the grid points yields

$$
\begin{array}{ll}
B_0(0) = B_n(1) = 1, & B_0(h) = B_n(1-h) = 1/6, \\
B_1(0) = B_{n-1}(1) = 0, & \\
B_1(h) = B_{n-1}(1-h) = 2/3, & B_1(2h) = B_{n-1}(1-2h) = 1/6, \\
B_j(x_j) = 2/3,\ B_j(x_{j\pm 1}) = 1/6 & \text{for } 2 \le j \le n-2.
\end{array}
\tag{4.12}
$$

One verifies that $y_j := \sum_{j=0}^{n} B_j(x_j) = 1$. Since the constant function $S = 1 \in V_n$ is interpolating, the unique solvability of the spline interpolation proves (4.11). Now we return to a general spline function $S = \sum_{i=0}^{n} y_i \Phi_{i,n}$. A representation by B-splines reads $S = \sum_{j=0}^{n} b_j B_j$. Note that $y_i = S(x_i) = \sum_{j=0}^{n} b_j B_j(x_i)$. Inserting the values from (4.12), we obtain

$$
y = Ab \qquad \text{with } A = \frac{1}{6}
\begin{bmatrix}
6 & & & & \\
1 & 4 & 1 & & \\
& \ddots & \ddots & \ddots & \\
& & 1 & 4 & 1 \\
& & & & 6
\end{bmatrix} b
$$

for the vectors $y = (y_i)_{i=0}^{n}$ and $b = (b_i)_{i=0}^{n}$. A can be written as $A = \frac{2}{3}[I + \frac{1}{2}C]$ with $\|C\|_\infty = 1$; i.e., A is strongly diagonal dominant and the inverse satisfies $\|A^{-1}\|_\infty \le 3$ because of

$$
A^{-1} = \frac{3}{2}[I + \frac{1}{2}C]^{-1} = \frac{3}{2} \sum_{\nu=0}^{\infty} 2^{-\nu} C^\nu.
$$

Using $b = A^{-1}y$, we derive from $S = \sum_{j=0}^{n} b_j B_j$ that

$$
|S(x)| = \left| \sum_{j=0}^{n} b_j B_j(x) \right| \underset{B_j \ge 0}{\le} \sum_{j=0}^{n} |b_j| B_j(x) \le \|b\|_\infty \sum_{j=0}^{n} B_j(x) \underset{(4.11)}{=} \|b\|_\infty
$$

for all $x \in J$, so that the stability estimate $\|S\|_\infty \le C_{\text{stab}} \|y\|_\infty$ is proved with $C_{\text{stab}} := 3$.

---

[18] The explicit polynomials are

$$
B_j = \frac{1}{6h^3}
\left\{
\begin{array}{ll}
\xi^3, & \xi = x - x_{j-2},\ x \in J_{j-1}, \\
h^3 + 3h^2\xi + 3h\xi^2 - 3\xi^3, & \xi = x - x_{j-1},\ x \in J_j, \\
h^3 + 3h^2\xi + 3h\xi^2 - 3\xi^3, & \xi = x_{j+1} - x,\ x \in J_{j+1}, \\
\xi^3, & \xi = x_{j+2} - x,\ x \in J_{j+2},
\end{array}
\right\}
\text{ for } 2 \le j \le n-2,
$$

$$
B_1 = \frac{1}{6h^3}
\left\{
\begin{array}{ll}
6h^2 x - 2x^3, & x \in J_1, \\
h^3 + 3h^2\xi + 3h\xi^2 - 3\xi^3, & \xi = 2h - x,\ x \in J_2, \\
\xi^3, & \xi = 3h - x,\ x \in J_3,
\end{array}
\right\}, \quad B_{n-1}(x) = B_1(1-x),
$$

$$
B_0 = \frac{1}{6h^3}
\left\{
\begin{array}{ll}
h^3 + 3h^2\xi + 3h\xi^2 - \xi^3, & \xi = h - x,\ x \in J_1, \\
\xi^3, & \xi = 2h - x,\ x \in J_2,
\end{array}
\right\}, \quad B_n(x) = B_0(1-x).
$$

Remark 4.15. *The previous results show that consistency is in conflict with stability. Polynomial interpolation has an increasing order of consistency, but suffers from instability (cf. Theorem 4.14). On the other hand, piecewise polynomial interpolation of bounded order is stable.*

4.9 From Point-wise Convergence to Operator-Norm Convergence

As already mentioned in §3.5 in the context of quadrature, only *point-wise convergence* $I_n(f) \to f$ ($f \in X$) can be expected, but not operator-norm convergence $\|I_n - id\| \to 0$. However, there are situations in which point-wise convergence can be converted into operator-norm convergence.

An operator $K : X \to Y$ is called *compact* if the image $B := \{Kf : \|f\|_X \leq 1\}$ is precompact (cf. page 44). The following theorem is formulated for an arbitrary, point-wise convergent sequence of operators $A_n : Y \to Z$.

Theorem 4.16. *Let X, Y, Z be Banach spaces, and $A, A_n \in \mathcal{L}(Y, Z)$. Suppose that point-wise convergence $A_n y \to A y$ holds for all $y \in Y$. Furthermore, let $K : X \to Y$ be compact. Then the products $P_n := A_n K$ converge with respect to the operator norm to $P := AK$; i.e., $\|P_n - P\| \to 0$.*

Proof. $M := \{Kx : \|x\|_X \leq 1\} \subset Y$ is precompact because of the compactness of K, so that we can apply Lemma 3.49:

$$\|P_n - P\| = \sup\{\|P_n x - Px\|_Z : x \in X, \|x\| \leq 1\}$$
$$= \sup\{\|A_n(Kx) - A(Kx)\|_Z : x \in X, \|x\| \leq 1\}$$
$$= \sup\{\|A_n y - Ay\|_Z : y \in M\} \underset{(3.29)}{\to} 0. \qquad \square$$

A typical example of a compact operator is the embedding

$$E : \left(C^\lambda([0,1]), \|\cdot\|_{C^\lambda([0,1])}\right) \to (C([0,1]), \|\cdot\|_\infty).$$

For integer $\lambda \in \mathbb{N}$, $C^\lambda([0,1])$ is the space of λ-times continuously differentiable functions, where the norm $\|\cdot\|_{C^\lambda([0,1])}$ is the maximum of all derivatives up to order λ. For $0 < \lambda < 1$, $C^\lambda([0,1])$ are the Hölder continuous functions with $\|\cdot\|_{C^\lambda([0,1])}$ explained in Footnote 9 on page 44. The embedding is the identity mapping: $E(f) = f$; however, the argument f and the image $E(f)$ are associated with different norms. As mentioned in the proof of Theorem 3.50, $E \in \mathcal{L}(C^\lambda([0,1]), C([0,1]))$ is compact.

In the case of $\lambda = 4$, estimate (4.10) already yields the operator-norm convergence $\|I_n - id\|_{C([0,1]) \leftarrow C^\lambda([0,1])} \leq C/n^4 \to 0$. To show a similar operator-norm convergence for $0 < \lambda < 1$, interpret $I_n - id$ as $(I_n - id)E : C^\lambda([0,1]) \to C([0,1])$. Applying Theorem 4.16 with $A = id$, $A_n = I_n$, and $K = E$, we obtain

$$\|I_n - id\|_{C([0,1]) \leftarrow C^\lambda([0,1])} \to 0.$$

4.10 Approximation

Often, interpolation is used as a simple tool to obtain an approximation; i.e., the interpolation condition (4.1) is not essential. Instead, we can directly ask for a best approximation $\Phi_n \in V_n$ of $f \in B$, where $V_n \subset B$ is an $(n + 1)$-dimensional subspace of a Banach space B with norm $\|\cdot\|$:

$$\|f - \Phi_n\| = \inf \{\|f - g\| : g \in V_n\}. \tag{4.13}$$

Using compactness arguments one obtains the existence of a minimiser Φ. If the space B is strictly convex,[19] the minimiser is unique (cf. [10]).

A prominent choice of V_n are the polynomials of degree $\leq n$, while $B = C([a, b])$ is equipped with the maximum norm. Polynomials satisfy the following *Haar condition*: any $0 \neq f \in V_n$ has at most n zeros (cf. Haar [7]). As a consequence, also in this case, the best approximation problem (4.13) has a unique solution. For the numerical solution of the best approximation the following equioscillation property is essential (cf. Chebyshev [2]):

Theorem 4.17. *Let $\varepsilon := f - \Phi_n$ be the error of the best approximation in (4.13). Then there are $n + 2$ points x_μ with $a \leq x_0 < x_1 < \ldots < x_{n+1} \leq b$ such that*

$$|\varepsilon(x_\mu)| = \|f - \Phi_n\| \quad and \quad \varepsilon(x_\mu) = -\varepsilon(x_{\mu+1}) \quad for\ 0 \leq \mu \leq n. \tag{4.14}$$

The second part of (4.14) describes $n + 1 = \dim(V_n)$ equations, which are used by the Remez algorithm to determine $\Phi_n \in V_n$ (cf. Remez [16]).

From (4.14) one concludes that there are n zeros $\xi_1 < \ldots < \xi_n$ of $\varepsilon = f - \Phi_n$; i.e., Φ_n can be regarded as an interpolation polynomial with these interpolation points. However note that the ξ_μ depend on the function f.

The mapping $f \mapsto \Phi_n$ is in general nonlinear. Below, when we consider Hilbert spaces, it will become a linear projection.

Since the set of polynomials is dense in $C([a, b])$ (cf. Theorem 3.28), the condition

$$V_0 \subset V_1 \subset \ldots \subset V_n \subset V_{n+1} \subset \ldots \text{ and } \overline{\bigcup_{n \in \mathbb{N}_0} V_n} = C([a, b]) \tag{4.15}$$

is satisfied. Condition (4.15) implies

$$\|f - \Phi_n\| \searrow 0 \quad \text{as } n \to \infty \quad \text{for } \Phi_n \text{ from (4.13)}. \tag{4.16}$$

Stability issues do not appear in this setting. One may consider the sequence $\{\|\Phi_n\| : n \in \mathbb{N}_0\}$, but (4.16) proves convergence $\|\Phi_n\| \to \|f\|$; i.e., the sequence must be uniformly bounded.

The approximation is simpler if B is a Hilbert space with scalar product $\langle \cdot, \cdot \rangle$. Then the best approximation from (4.13) is obtained by means of the orthogonal

---

[19] B is strictly convex if $\|f\| = \|g\| = 1$ and $f \neq g$ imply $\|f + g\| < 2$.

projection[20] $\Pi_n \in \mathcal{L}(B, B)$ onto V_n:

$$\Phi_n = \Pi_n f.$$

Given any orthonormal basis $\{\phi_\mu : 0 \leq \mu \leq n\}$ of V_n, the solution has the explicit representation

$$\Phi_n = \sum_{\mu=0}^{n} \langle f, \phi_\mu \rangle \, \phi_\mu. \qquad (4.17)$$

The standard example is the Fourier approximation of 2π periodic real functions in $[-\pi, \pi]$. The L^2 scalar product is $\langle f, g \rangle = \int_{-\pi}^{\pi} fg \mathrm{d}x$. Let n be even. V_n is spanned by the orthonormal basis functions

$$\left\{ \frac{1}{\sqrt{2\pi}}, \frac{\cos(mx)}{\sqrt{\pi}}, \frac{\sin(mx)}{\sqrt{\pi}} : 1 \leq m \leq n/2 \right\}.$$

At first glance there is no stability problem to be discussed, since the operator norm of orthogonal projections equals one: $\|\Pi_n\|_{L^2 \leftarrow L^2} = 1$. However, if we consider the operator norm $\|\Pi_n\|_{B \leftarrow B}$ for another Banach space, (in)stability comes into play.

Let Π_n be the Fourier projection from above and choose the Banach space $B = C_{2\pi} := \{f \in C([-\pi, \pi]) : f(-\pi) = f(\pi)\}$ equipped with the maximum norm $\|\cdot\|_\infty$. We ask for the behaviour of $\|\Pi_n\|_\infty$, where now $\|\cdot\|_\infty = \|\cdot\|_{C_{2\pi} \leftarrow C_{2\pi}}$ denotes the operator norm. The mapping (4.17) can be reformulated by means of the Dirichlet kernel,

$$(\Phi_n f)(x) = \frac{1}{\pi} \int_0^\pi \frac{\sin(2n+1)y}{\sin(y)} [f(x+2y) + f(x-2y)] \mathrm{d}y.$$

From this representation we infer that

$$\|\Pi_n\|_\infty = \frac{1}{\pi} \int_0^\pi \left| \frac{\sin(2n+1)y}{\sin(y)} \right| \mathrm{d}y.$$

Lower and upper bounds of this integral are

$$\frac{4}{\pi^2} \log(n+1) \leq \|\Pi_n\|_\infty \leq 1 + \log(2n+1).$$

This shows that the Fourier projection is unstable with respect to the maximum norm. The negation of the uniform boundedness theorem 3.38 together with $\|\Pi_n\|_\infty \to \infty$ implies the well-known fact that uniform convergence $\Pi_n f \to f$ cannot hold for any $f \in C_{2\pi}$.

The orthogonal Fourier projection Π_n is the best choice for the Hilbert space $L^2([-\pi, \pi])$. For $C_{2\pi}$ one may choose another projection P_n from $C_{2\pi}$ onto V_n.

---

[20] That means (i) $\Pi_n \Pi_n = \Pi_n$ (projection property) and (ii) Π_n is selfadjoint: $\langle \Pi_n f, g \rangle = \langle f, \Pi_n g \rangle$ for all $f, g \in B$.

This, however, can only lead to larger norms $\|P_n\|_\infty$ due to the following result of Cheney et al. [3].

Theorem 4.18. *The Fourier projection Π_n is the unique minimiser of*

$$\min\{\|P_n\|_\infty : P_n \in \mathcal{L}(C_{2\pi}, C_{2\pi}) \text{ projection onto } V_n\}.$$

References

1. Bernstein, S.N.: Sur la limitation des valeurs d'un polynôme $p(x)$ de degré n sur tout un segment par ses valeurs en $(n + 1)$ points du segment. Izv. Akad. Nauk SSSR **8**, 1025–1050 (1931)
2. Chebyshev, P.L.: Sur les questions de minima qui se rattachent à la représentation approximative des fonctions. In: Oeuvres, Vol. I, pp. 641–644. St. Petersburg (1899)
3. Cheney, E.W., Hobby, C.R., Morris, P.D., Schurer, F., Wulbert, D.E.: On the minimal property of the Fourier projection. Trans. Am. Math. Soc. **143**, 249–258 (1969)
4. Erdös, P.: Problems and results on the theory of interpolation. Acta Math. Acad. Sci. Hungar. **12**, 235–244 (1961)
5. Euler, L.: De progressionibus harmonicis observationes. Commentarii academiae scientiarum imperialis Petropolitanae **7**, 150–161 (1740)
6. Faber, G.: Über die interpolatorische Darstellung stetiger Funktionen. Jber. d. Dt. Math.-Verein. **23**, 190–210 (1914)
7. Haar, A.: Die Minkowskische Geometrie und die Annäherung an stetige Funktionen. Math. Ann. **78**, 294–311 (1918)
8. Hackbusch, W.: Hierarchische Matrizen - Algorithmen und Analysis. Springer, Berlin (2009)
9. Hackbusch, W.: Tensor spaces and numerical tensor calculus, *Springer Series in Computational Mathematics*, Vol. 42. Springer, Berlin (2012)
10. Meinardus, G.: Approximation of functions: theory and numerical methods. Springer, New York (1967)
11. Mills, T.M., Smith, S.J.: The Lebesgue constant for Lagrange interpolation on equidistant nodes. Numer. Math. pp. 111–115 (1992)
12. Natanson, I.P.: Konstruktive Funktionentheorie. Akademie-Verlag, Berlin (1955)
13. Natanson, I.P.: Constructive function theory, Vol. III. Frederick Ungar Publ., New York (1965)
14. Plato, R.: Concise Numerical Mathematics. AMS, Providence (2003)
15. Quarteroni, A., Sacco, R., Saleri, F.: Numerical Mathematics, 2nd ed. Springer, Berlin (2007)
16. Remez, E.J.: Sur un procédé convergent d'approximations successives pour déterminer les polynômes d'approximation. Compt. Rend. Acad. Sc. **198**, 2063–2065 (1934)
17. Rivlin, T.J.: Chebyshev Polynomials. Wiley, New York (1990)
18. Schönhage, A.: Fehlerfortpflanzung bei Interpolation. Numer. Math. **3**, 62–71 (1961)
19. Stoer, J., Bulirsch, R.: Introduction to Numerical Analysis. North-Holland, Amsterdam (1980)
20. Trefethen, L.N., Weideman, J.A.C.: Two results on polynomial interpolation in equally spaced points. J. Approx. Theory **65**, 247–260 (1991)
21. Turetskii, A.H.: The bounding of polynomials prescribed at equally distributed points (Russian). Proc. Pedag. Inst. Vitebsk **3**, 117–127 (1940)

Chapter 5
Ordinary Differential Equations

The numerical treatment of ordinary differential equations is a field whose scope has broadened quite a bit over the last fifty years. In particular, a whole spectrum of different stability conditions has developed. Since this chapter is not the place to present all details, we concentrate on the most basic concept of stability. As a side-product, it will lead us to the power bounded matrices, which is a class of matrices with certain stability properties. More details about ordinary differential equations can be found, e.g., in [20], [23], [4, 3], [5, §§5-6], and in the two volumes [8], [9].

5.1 Initial-Value Problem

5.1.1 Setting of the Problem

Let $f : \mathbb{R} \times \mathbb{R} \to \mathbb{R}$ be a continuous function.[1] In what follows we are looking for continuously differentiable functions $y(x)$ satisfying the *ordinary differential equation*

$$y'(x) = f(x, y(x)). \tag{5.1a}$$

The *initial-value problem* requires finding a solution y of (5.1a) which, in addition, satisfies

$$y(x_0) = y_0 \tag{5.1b}$$

for a given 'initial value' y_0.

Usually one is looking for the solution y at $x \geq x_0$, either in a finite[2] interval $I := [x_0, x_E]$ or in the unbounded interval $I := [x_0, \infty)$. Correspondingly, f needs to be defined on $I \times \mathbb{R}$.

---

[1] In the case of a *system* of differential equations, f is defined in $\mathbb{R} \times \mathbb{R}^n$ and the solution $y \in C^1(\mathbb{R}, \mathbb{R}^n)$ is vector-valued. For our considerations it is sufficient to study the scalar case $n = 1$.

[2] However, it may happen that the solution exists only on a smaller interval $[x_0, x_S] \subset [x_0, x_E]$.

W. Hackbusch, *The Concept of Stability in Numerical Mathematics*,
Springer Series in Computational Mathematics 45, DOI 10.1007/978-3-642-39386-0_5,
© Springer-Verlag Berlin Heidelberg 2014

If f is only continuous, a solution of (5.1a,b) exists due to Peano's theorem (at least, in the neighbourhood $[x_0, x_0 + \varepsilon)$ for some $\varepsilon > 0$); however, there may be more than one solution. Uniqueness is ensured by the assumption that f is Lipschitz continuous with respect to the second argument (cf. Lipschitz [16, pp. 500ff]. For simplicity, we formulate the global *Lipschitz continuity*:

$$|f(x, y_1) - f(x, y_2)| \le L |y_1 - y_2| \qquad \text{for all } x \in I, \ y_1 - y_2 \in \mathbb{R}. \qquad (5.2)$$

The Lipschitz constant L will appear in later analysis. Unique solvability will be stated in Corollary 5.10.

5.1.2 One-Step Methods

We choose a fixed step size[3] $h > 0$. The corresponding grid points are

$$x_i := x_0 + ih \qquad (i \in \mathbb{N}_0, x_i \in I).$$

Next, we define approximations η_i of $y(x_i)$. The notation η_i assumes that an underlying step size h is defined. If necessary, we write $\eta(x_0 + ih; h)$ instead of η_i. Note that $\eta(x; h)$ is defined only for grid points $x = x_0 + ih$ ($i \in \mathbb{N}_0$). The desired property is $\eta(x; h) \approx y(x)$, where the error should tend to zero as $h \to 0$.

Because of the given initial value y_0, we start the computation with

$$\eta_0 := y_0. \qquad (5.3)$$

The prototype of the one-step methods is the *Euler method*, which starts with (5.3) and defines recursively

$$\eta_{i+1} := \eta_i + hf(x_i, \eta_i). \qquad (5.4)$$

Exercise 5.1. *Consider the differential equation $y' = ay$ (i.e., $f(x, y) = ay$) with the initial value $y_0 = 1$ at $x_0 = 0$ (its exact solution is $y(x) = e^{ax}$). Determine the solution of (5.4). Does $y(x) - \eta(x; h) \to 0$ hold for $h := x/n > 0$, when $n \to \infty$ and $h \to 0$ for fixed $nh = x$?*

For other one-step methods, one replaces the right-hand side in (5.4) by a more general expression $h\phi(x_i, \eta_i, h; f)$. Here, the last argument f means that inside of ϕ the function f can be used for arbitrary evaluations. The Euler method corresponds to $\phi(x_i, \eta_i, h; f) = f(x_i, \eta_i)$.

Definition 5.2. *A general explicit one-step method has the form*

$$\eta_{i+1} := \eta_i + h\phi(x_i, \eta_i, h; f). \qquad (5.5)$$

---

[3] In practical implementations one has to admit varying step widths $h_i = x_{i+1} - x_i$. Usually, these are chosen adaptively.

Often, the evaluation of ϕ is performed in several partial steps. For instance, the *Heun method* [10] uses the intermediate step $\eta_{i+1/2}$:

$$\eta_{i+1/2} := \eta_i + \frac{h}{2}f(x_i, \eta_i), \quad \eta_{i+1} := \eta_i + hf(x_i + \frac{h}{2}, \eta_{i+1/2}).$$

These equations yield $\phi(x_i, \eta_i, h; f) := f(x_i + \frac{h}{2}, \eta_i + \frac{h}{2}f(x_i, \eta_i))$. The classical *Runge–Kutta* method uses four intermediate steps (cf. Runge [19], Kutta [15]; history in [1]; for a modern description see, e.g., [8, §II]).

5.1.3 Multistep Methods

The term 'one-step method' refers to the fact that (x_{i+1}, η_{i+1}) is determined only by (x_i, η_i). The past values η_j, $j < i$, do not enter into the algorithm.

On the other hand, since besides η_i the values $\eta_{i-r}, \eta_{i-r+1}, \ldots, \eta_{i-1}$ are available (r is a fixed natural number), one may ask whether one can use these data. Indeed, having more free parameters, one can try to increase the order of the method. This leads to the r-step method, which is of the form

$$\eta_{j+r} := -\sum_{\nu=0}^{r-1} \alpha_\nu \eta_{j+\nu} + h\phi(x_j, \eta_{j+r-1}, \ldots, \eta_j, h; f) \tag{5.6}$$

(more precisely, this is the explicit form of a multistep method) with the additional parameters $\alpha_0, \ldots, \alpha_{r-1} \in \mathbb{R}$ for $j = 0, \ldots, r - 1$. As we shall see,

$$\sum_{\nu=0}^{r-1} \alpha_\nu = -1 \tag{5.7}$$

describes a first consistency condition.

Remark 5.3. *Because of (5.7), a multistep method (5.6) with $r = 1$ coincides with the one-step method (5.5).*

Remark 5.4. *In the case of $r \geq 2$, the multistep methods (5.6) can only be used for the computation of η_i for $i \geq r$. The computation of $\eta_1, \eta_2, \ldots, \eta_{r-1}$ must be defined in another way (e.g., by a one-step method).*

An example of a two-step method is the *midpoint formula*

$$\eta_{j+2} = \eta_j + 2hf(x_{j+1}, \eta_{j+1}); \tag{5.8}$$

i.e., $r = 2$, $\alpha_0 = -1$, $\alpha_1 = 0$, $\phi(x_j, \eta_{j+1}, \eta_j, h; f) = 2f(x_j + h, \eta_{j+1})$.

A rather dubious proposal is the extrapolation

$$\eta_{j+2} = 2\eta_{j+1} - \eta_j; \tag{5.9}$$

i.e., $r = 2$, $\alpha_0 = -2$, $\alpha_1 = 1$, $\phi(x_j, \eta_{j+1}, \eta_j, h; f) = 0$. Even without any quantitative error analysis one can see that the latter two-step method cannot be successful. Since (5.9) does not depend on f, it produces the same linear function $\eta_j = \eta_0 + j(\eta_1 - \eta_0)$ for all differential equations.

5.2 Fixed-Point Theorem and Recursive Inequalities

We now provide some technical tools required in this chapter.

Theorem 5.5 (Banach's fixed-point theorem). *Let X be a Banach space and $\Psi : X \to X$ a contraction mapping; i.e., there is a constant $L_\Psi \in [0, 1)$ such that*

$$\|\Psi(x') - \Psi(x'')\|_X \le L_\Psi \|x' - x''\|_X \qquad \text{for all } x', x'' \in X. \tag{5.10a}$$

Then the fixed-point equation $x = \Psi(x)$ has exactly one solution, and for all starting values $x_0 \in X$ the fixed-point iteration $x_{n+1} = \Psi(x_n)$ converges to this solution. Furthermore, the error of the n-th iterant can be estimated by

$$\|x - x_n\|_X \le \frac{(L_\Psi)^n}{1 - L_\Psi} \|x_1 - x_0\|_X \qquad \text{for all } n \in \mathbb{N}_0. \tag{5.10b}$$

Proof. (i) Let x', x'' be two solutions of the fixed-point equation; i.e., $x' = \Psi(x')$ and $x'' = \Psi(x'')$. Exploiting the contraction property with $L = L_\Psi$, we get

$$\|x' - x''\|_X = \|\Psi(x') - \Psi(x'')\|_X \le L \|x' - x''\|_X .$$

From $L < 1$ we conclude that $\|x' - x''\|_X = 0$; i.e., uniqueness $x' = x''$ is proved.
(ii) The iterants x_n of the fixed-point iteration satisfy the inequality

$$\|x_{n+1} - x_n\|_X = \|\Psi(x_n) - \Psi(x_{n-1})\|_X \le L \|x_n - x_{n-1}\|_X$$

and thereby $\|x_{n+1} - x_n\|_X \le L^n \|x_1 - x_0\|_X$. For arbitrary $n > m$, the multiple triangle inequality yields the estimate

$$\|x_n - x_m\|_X \le \sum_{j=m+1}^{n} \|x_j - x_{j-1}\|_X \le \sum_{j=m+1}^{n} L^{j-1} \|x_1 - x_0\|_X$$

$$\le \sum_{j=m}^{\infty} L^j \|x_1 - x_0\|_X = \frac{L^m}{1 - L} \|x_1 - x_0\|_X ; \tag{5.10c}$$

i.e., $\{x_n\}$ is a Cauchy sequence. Since X is a Banach space and therefore complete, a limit $x^* = \lim x_n$ exists. Because of the continuity of the function Ψ, the limit in $x_{n+1} = \Psi(x_n)$ yields $x^* = \Psi(x^*)$; i.e., x^* is a fixed-point solution.
  Inequality (5.10b) follows from (5.10c) for $n \to \infty$. $\square$

In the following analysis, recursive inequalities of the following form will appear:

$$a_{\nu+1} \leq (1 + hL)\, a_\nu + h^k B \qquad \text{for all } \nu \geq 0, \text{where } L, B, h, k, a_0 \geq 0. \quad (5.11)$$

The meaning of the parameters is: L Lipschitz constant, h step size, and k local consistency order.

Lemma 5.6. *Any solution of the inequalities (5.11) satisfies the estimate*

$$a_\nu \leq e^{\nu hL} a_0 + h^{k-1} B \cdot \left\{ \begin{array}{ll} \nu h & \text{for } L = 0 \\[2mm] \dfrac{e^{\nu hL} - 1}{L} & \text{for } L > 0 \end{array} \right\} \qquad (\nu \in \mathbb{N}_0).$$

Proof. We pose the following induction hypothesis:

$$a_\nu \leq A_\nu := \sum_{\mu=0}^{\nu-1} (1 + hL)^\mu\, h^k B + (1 + hL)^\nu\, a_0. \qquad (5.12)$$

The start of the induction is given by $a_0 \leq A_0 = a_0$. For the induction step $\nu \mapsto \nu + 1$ insert $a_\nu \leq A_\nu$ into (5.11):

$$a_{\nu+1} \leq (1 + hL)\, A_\nu + h^k B = \sum_{\mu=1}^{\nu} (1 + hL)^\mu\, h^k B + (1 + hL)^{\nu+1}\, a_0 + h^k B$$

$$= \sum_{\mu=0}^{\nu} (1 + hL)^\mu\, h^k B + (1 + hL)^{\nu+1}\, a_0 = A_{\nu+1}.$$

Exercise 3.24a shows that $(1 + hL)^\nu \leq \left(e^{hL}\right)^\nu = e^{hL\nu}$. For $L > 0$, the geometric sum yields the value

$$h^k B \sum_{\mu=0}^{\nu-1} (1 + hL)^\mu = h^k B \frac{(1 + hL)^\nu - 1}{(1 + hL) - 1} = h^{k-1} \frac{B}{L} \left[(1 + hL)^\nu - 1 \right]$$

$$\leq h^{k-1} \frac{B}{L} \left[e^{hL\nu} - 1 \right].$$

Therefore, A_ν from (5.12) can be estimated by $A_\nu \leq h^{k-1} \frac{B}{L} \left[e^{hL\nu} - 1 \right] + e^{hL\nu} a_0$. The particular case $L = 0$ can be treated separately or obtained from $L > 0$ by performing the limit $L \to 0$. $\square$

Exercise 5.7. *Prove that any solution φ of the inequality $\varphi(x) \leq \varphi_0 + L \int_{x_0}^{x} \varphi(t)\mathrm{d}t$ is bounded by Φ; i.e., $\varphi(x) \leq \Phi(x)$, where Φ is the solution of the integral equation*

$$\Phi(x) = \varphi_0 + L \int_{x_0}^{x} \Phi(t)\mathrm{d}t.$$

Hint: (a) Define $\Psi(\Phi)$ by $\Psi(\Phi)(x) := \varphi_0 + L \int_{x_0}^{x} \Phi(t)\mathrm{d}t$. The integral equation is the fixed-point equation $\Phi = \Psi(\Phi)$ for $\Phi \in C(I)$, $I = [x_0, x_E]$. Show that Ψ is a contraction with respect to the norm

$$\|\psi\| := \max\{|\psi(t)| \exp(-2L(t - x_0)) : t \in I\}$$

with contraction number $L_{\Psi} = \frac{1}{2}$ (independently of the length of the interval I, including $I = [x_0, \infty)$).
(b) Apply the fixed-point iteration $\Phi_{n+1} := \Psi(\Phi_n)$ with $\Phi_0 := \varphi$ and show $\Phi_n \geq \varphi$ for all $n \geq 0$.

Lemma 5.8. Let X and Y be Banach spaces. The operators $S, T \in \mathcal{L}(X, Y)$ are supposed to satisfy

$$T^{-1} \in \mathcal{L}(Y, X) \quad \text{and} \quad \|S - T\|_{Y \leftarrow X} \|T^{-1}\|_{X \leftarrow Y} < 1.$$

Then also the inverse $S^{-1} \in \mathcal{L}(Y, X)$ exists and satisfies

$$\|S^{-1}\|_{X \leftarrow Y} \leq \frac{\|T^{-1}\|_{X \leftarrow Y}}{1 - \|S - T\|_{Y \leftarrow X} \|T^{-1}\|_{X \leftarrow Y}}.$$

Proof. Fix $y \in X$. The mapping $\Phi(x) := T^{-1}(T - S)x + T^{-1}y$ is contracting with contraction rate $q = \|T^{-1}\|_{X \leftarrow Y} \|S - T\|_{Y \leftarrow X} < 1$. Banach's fixed-point Theorem 5.5 states the unique solvability of $\Phi(x) = x$, which implies that $Tx = T\Phi(x) = (T - S)x + y$ and $Sx = y$. Hence the inverse S^{-1} exists. The fixed-point iteration with starting value x_0 produces $x_1 = T^{-1}y$. Estimate (5.10b) with $n = 0$ yields

$$\|S^{-1}y\| = \|S^{-1}y - x_0\| = \|x - x_0\| \leq \frac{1}{1 - q}\|x_1 - x_0\| = \frac{1}{1 - q}\|T^{-1}y\|,$$

which leads to the desired inequality.   □

5.3 Well-Conditioning of the Initial-Value Problem

Before we start with the numerical solution, we should check whether the initial-value problem, i.e., the mapping $(y_0, f) \mapsto y$, is well-conditioned. According to §2.4.1.2, the amplification of a perturbation of the input data is to be investigated. In the present case, one can perturb the initial value y_0 as well as the function f. The first case is analysed below, the second one in Theorem 5.12.

Theorem 5.9. Let $y_1, y_2 \in C^1(I)$ be two solutions of the differential equation (5.1a) with the initial values

$$y_1(x_0) = y_{0,1} \quad \text{and} \quad y_2(x_0) = y_{0,2} \quad \text{respectively.}$$

Assume that $f \in C(I \times \mathbb{R})$ satisfies (5.2). Then the following estimate[4] holds in I:

$$|y_1(x) - y_2(x)| \leq |y_{0,1} - y_{0,2}| e^{L(x - x_0)} \qquad \text{with } L \text{ from (5.2).} \qquad (5.13)$$

---

[4] By definition of I in §5.1, $x \geq x_0$ holds. Otherwise, $e^{L(x - x_0)}$ is to be replaced by $e^{L|x - x_0|}$.

Proof. $y_i(x) = y_{0,i} + \int_{x_0}^x f(t, y_i(t)) dt$ holds for $i = 1, 2$, so that

$$|y_1(x) - y_2(x)| = \left| y_{0,1} - y_{0,2} + \int_{x_0}^x [f(t, y_1(t)) - f(t, y_2(t))] dt \right|$$

$$\leq |y_{0,1} - y_{0,2}| + \int_{x_0}^x |f(t, y_1(t)) - f(t, y_2(t))| dt \leq \quad (5.2)$$

$$\leq |y_{0,1} - y_{0,2}| + \int_{x_0}^x L |y_1(t) - y_2(t)| dt.$$

The function $\Phi(x) := |y_{0,1} - y_{0,2}| e^{L(x-x_0)}$ satisfies the equation

$$\Phi(x) = |y_{0,1} - y_{0,2}| + \int_{x_0}^x L\Phi(t) dt \qquad \text{on } I.$$

Hence Exercise 5.7 proves $|y_1(x) - y_2(x)| \leq \Phi(x)$; i.e., (5.13). This ends the proof of Theorem 5.9. ☐

Corollary 5.10. *Assumption (5.2) ensures uniqueness of the solution of the initial-value problem (5.1a,b).*

Proof. If $y_1, y_2 \in C^1(I)$ are two solutions, Theorem 5.9 yields

$$|y_1(x) - y_2(x)| \leq |y_{0,1} - y_{0,2}| \exp(L(x - x_0)) \underset{y_{0,1} = y_{0,2}}{=} 0,$$

hence $y_1 = y_2$ on I. ☐

One may denote the solution of the initial-value problem (5.1a,b) by $y(x; y_0)$ with the initial value y_0 as second argument. Then (5.13) states that $y(\cdot; \cdot)$ as well as $f(\cdot, \cdot)$ are Lipschitz continuous with respect to the second argument. This statement can be generalised.

Exercise 5.11. *If $f(\cdot, \cdot)$ is k-times continuously differentiable with respect to the second argument y, also the solution $y(\cdot; \cdot)$ does so with respect to y_0.*

A perturbation of the right-hand side f in (5.1a) is studied next.

Theorem 5.12. *Let y and $\tilde{y}$ be solutions of $y' = f(x, y)$ and $\tilde{y}' = \tilde{f}(x, \tilde{y})$, respectively, with coinciding initial values $y(x_0) = \tilde{y}(x_0) = y_0$. Only f (not $\tilde{f}$) has to fulfil the Lipschitz condition (5.2), while*

$$\left| f(x, y) - \tilde{f}(x, y) \right| \leq \varepsilon \qquad \text{for all } x \in I, y \in \mathbb{R}.$$

Then

$$|y(x) - \tilde{y}(x)| \leq \left\{ \begin{array}{ll} \frac{\varepsilon}{L} \left(e^{L(x-x_0)} - 1 \right) & \text{if } L > 0 \\ \varepsilon(x - x_0) & \text{if } L = 0 \end{array} \right\}, \qquad (L \text{ from } (5.2)).$$

Proof. Set $\delta(x) := |y(x) - \tilde{y}(x)|$ and note that

$$
\begin{aligned}
\delta(x) &= \left| \int_{x_0}^{x} \left[f(\xi, y(\xi)) - \tilde{f}(\xi, \tilde{y}(\xi)) \right] \mathrm{d}\xi \right| \leq \int_{x_0}^{x} \left| \left[f(\xi, y(\xi)) - \tilde{f}(\xi, \tilde{y}(\xi)) \right] \right| \mathrm{d}\xi \\
&= \int_{x_0}^{x} \left| f(\xi, y(\xi)) - f(\xi, \tilde{y}(\xi)) + f(\xi, \tilde{y}(\xi)) - \tilde{f}(\xi, y(\xi)) \right| \mathrm{d}\xi \\
&\leq \int_{x_0}^{x} |f(\xi, y(\xi)) - f(\xi, \tilde{y}(\xi))| \, \mathrm{d}\xi + \int_{x_0}^{x} \left| f(\xi, \tilde{y}(\xi)) - \tilde{f}(\xi, y(\xi)) \right| \mathrm{d}\xi \\
&\leq \int_{x_0}^{x} L\delta(\xi) \mathrm{d}\xi + (x - x_0)\varepsilon.
\end{aligned}
$$

A majorant of δ is the solution d of $d(x) = \int_{x_0}^{x} L d(\xi) \mathrm{d}\xi + (x - x_0)\varepsilon$. In the case of $L > 0$, the solution is $d(x) = \frac{\varepsilon}{L} \left(e^{L(x-x_0)} - 1 \right)$. $\square$

5.4 Analysis of One-Step Methods

In §5.4.1 we explain why for our analysis in §§5.4.3–5.4.5 it is sufficient to study *explicit* one-step methods. In §5.4.2 we discuss the Lipschitz continuity of ϕ.

5.4.1 Implicit Methods

Definition 5.13. *A general implicit one-step method has the form*

$$
\eta_{i+1} := \eta_i + h\phi(x_i, \eta_i, \eta_{i+1}, h; f). \tag{5.14}
$$

An example of (5.14) is the *implicit Euler method*, where

$$
\phi(x_i, \eta_i, \eta_{i+1}, h; f) = f(x_i + h, \eta_{i+1}). \tag{5.15}
$$

Next we assume that ϕ is defined for $x_i \in I$, $\eta_i, \eta_{i+1} \in \mathbb{R}$, and sufficiently small h ($0 < h \leq h_0$).

Exercise 5.14. *Let ϕ from (5.14) be Lipschitz continuous with respect to η_{i+1}:*

$$
|\phi(x_i, \eta_i, \eta_{i+1}, h; f) - \phi(x_i, \eta_i, \hat{\eta}_{i+1}, h; f)| \leq L \, |\eta_{i+1} - \hat{\eta}_{i+1}|.
$$

Show that the fixed-point equation (5.14) is uniquely solvable, if $h < 1/L$.

Assuming unique solvability of (5.14), a function $\eta_{i+1} = \Psi(x_i, \eta_i, h; f)$ exists by the implicit function theorem. Inserting $\eta_{i+1} = \Psi(x_i, \eta_i, h; f)$ into the third argument of $\phi(x_i, \eta_i, \eta_{i+1}, h; f)$, one obtains formally an explicit one-step method (5.5) with

$$\hat{\phi}(x_i, \eta_i, h; f) := \phi(x_i, \eta_i, \Psi(x_i, \eta_i, h; f), h; f).$$

Hence, for theoretical considerations we may restrict ourselves to the explicit case (5.5). The only additional requirement is that one has to restrict the step sizes to $0 < h \le h_0$ with sufficiently small h_0. For the theoretical analysis the latter condition is of no consequence, since we study the limit $h \to 0$.

5.4.2 Lipschitz Continuity of ϕ

Analogous to the Lipschitz condition (5.2) we shall need Lipschitz continuity of ϕ:

$$|\phi(x_i, \eta', h; f) - \phi(x_i, \eta'', h; f)| \le L_\phi |\eta' - \eta''| \quad \text{for all} \begin{cases} x_i \in I, \\ \eta', \eta'' \in \mathbb{R}, \\ h \le h_0. \end{cases} \quad (5.16)$$

Since ϕ is defined implicitly via f, the Lipschitz property of ϕ is inherited from the Lipschitz continuity of f. Therefore, we have always to assume that f satisfies (5.2).

Exercise 5.15. *(a) Prove for the Euler and Heun methods that (5.2) implies (5.16). (b) According to §5.4.1, the implicit Euler method (5.15) leads to an explicit method with $\hat{\phi}(x_i, \eta_i, h; f)$. For sufficiently small h, prove Lipschitz continuity of $\hat{\phi}$.*

5.4.3 Consistency

We now motivate the consistency condition. Replacing the discrete solution η_i of $\eta_{i+1} := \eta_i + h\phi(x_i, \eta_i, h; f)$ by the exact solution $y(x_i)$ of the differential equation, we obtain the so-called *local discretisation error* τ by

$$y(x_{i+1}) = y(x_i) + h\left[\phi(x_i, y(x_i), h; f) + \tau(x_i, y(x_i); h)\right].$$

For the explicit definition of $\tau(\xi, \eta; h)$ fix $\xi \in I$ and $\eta \in \mathbb{R}$, and let $Y(\cdot\,; \xi, \eta)$ be the solution of (5.1a) with initial value condition

$$Y(\xi; \xi, \eta) = \eta \qquad \text{at } x = \xi \text{ (not at } x = x_0).$$

Then

$$\tau(\xi, \eta; h) := \frac{Y(\xi + h; \xi, \eta) - \eta}{h} - \phi(\xi, \eta, h; f) \qquad (5.17)$$

defines the local discretisation error at $(\xi, \eta; h)$.

Obviously, we may expect that the one-step method (5.5) is better the smaller τ is. Note that $\tau = 0$ leads to the ideal result $\eta_i = y(x_i)$.

Definition 5.16. *(a) The one-step method characterised by ϕ is called* consistent *if*

$$\sup_{x \in I} |\tau(x, y(x); h)| \to 0 \qquad (h \to 0). \tag{5.18}$$

Here y is the solution of (5.1a,b). The argument $f \in C(I \times \mathbb{R})$ of ϕ must satisfy (5.2). (b) Furthermore, ϕ is called consistent of order p *if $\tau(x, y(x); h) = \mathcal{O}(h^p)$ holds uniformly for $h \to 0$ on $x \in I$ for all sufficiently smooth f.*

Assuming f to be sufficiently smooth,[5] one performs the Taylor expansion of $\frac{1}{h}[y(x + h; x, \eta) - \eta]$ and uses

$$y(x + h; x, \eta) = y(x; x, \eta) + hy'(x; x, \eta) + o(h) = \eta + hf(x, \eta) + o(h).$$

Hence (5.18) implies the condition $\phi(x, \eta, h; f) \to f(x, \eta)$. One easily checks that this condition is satisfied for the methods of Euler and Heun.

However, the trivial one-step method $\eta_{i+1} := \eta_i$ (i.e., $\phi = 0$) leads, in general, to $\tau(x, \eta; h) = \mathcal{O}(1)$ and is not consistent.

5.4.4 Convergence

We recall the notation $\eta_i = \eta(x_i, h)$. The desired property is $\eta(x, h) \approx y(x)$. Concerning the limit $h \to 0$, we restrict ourselves tacitly to (a subsequence of) $h_n := (x - x_0)/n$, since then $x = nh_n$ belongs to the grid on which $\eta(\cdot, h_n)$ is defined.

Definition 5.17 (convergence). *A one-step method is called* convergent *if for all Lipschitz continuous f and all $x \in I$*

$$\lim_{h \to 0} \eta(x, h) = y(x) \qquad (y \text{ solution of } (5.1a,b))$$

holds. A one-step method has convergence order p *if $\eta(x, h) = y(x) + \mathcal{O}(h^p)$ for sufficiently smooth f.*

5.4.5 Stability

Consistency controls the error generated in the i-th step from x_i to x_{i+1} under the assumption that η_i is the exact starting value. At the start, $\eta_0 = y_0$ is indeed exact, so that according to condition (5.18) the error $\varepsilon_1 := \eta_1 - y_1$ is $o(h)$ or $\mathcal{O}(h^{p+1})$, respectively.

During the steps for $i \geq 1$ the consistency error, e.g., arising at x_1, is transported into $\eta_2, \eta_3, \ldots$ Since the computation proceeds up to $x = x_n$, one has to perform

---

[5] Without p-fold continuous differentiability of f one cannot verify $\tau(x, \eta; h) = \mathcal{O}(h^p)$.

$n = \mathcal{O}(1/h)$ steps. If the error would be amplified in each step by a factor $c > 1$ (c independently of h), η_n had an error $\mathcal{O}(c^n) = \mathcal{O}(c^{1/h})$. Obviously, such an error would explode exponentially as $h \to 0$. In addition, not only can the consistency error ε_1 be amplified, but also can all consistency errors ε_i at the later grid points x_i.

Next we state that—thanks to Lipschitz condition (5.16)—the errors are under control.[6]

Lemma 5.18 (stability of one-step methods). *Assume that the Lipschitz condition (5.16) holds with constant L_ϕ and the local discretisation error is bounded by $|\tau(x_i, y(x_i); h)| \le T_h$ (cf. (5.17)). Then the global discretisation error is bounded by*

$$|\eta(x,h) - y(x)| \le T_h \frac{e^{(x-x_0)L_\phi} - 1}{L_\phi}. \tag{5.19}$$

Proof. $\delta_i := |\eta_i - y(x_i)|$ is the *global error*. The local discretisation error is denoted by $\tau_i = \tau(x_i, y(x_i); h)$. Starting with $\delta_0 = 0$, we obtain the recursion formula

$$\begin{aligned}
\delta_{i+1} &= |\eta_{i+1} - y(x_{i+1})| = |\eta_i + h\phi(x_i, \eta_i, h; f) - y(x_{i+1})| \\
&= \left| \eta_i - y(x_i) - h\left[\frac{y(x_{i+1}) - y(x_i)}{h} - \phi(x_i, \eta_i, h; f) \right] \right| \\
&= \left| \eta_i - y(x_i) - h\left[\frac{y(x_{i+1}) - y(x_i)}{h} - \phi(x_i, y(x_i), h; f) \right] \right. \\
&\qquad \left. + h\left[\phi(x_i, \eta_i, h; f) - \phi(x_i, y(x_i), h; f) \right] \right| \\
&\le |\eta_i - y(x_i)| + h\left| \frac{y(x_{i+1}) - y(x_i)}{h} - \phi(x_i, y(x_i), h; f) \right| \\
&\qquad + h\left| \phi(x_i, \eta_i, h; f) - \phi(x_i, y(x_i), h; f) \right| \\
&\le \delta_i + h|\tau_i| + hL_\phi \delta_i = (1 + hL_\phi)\delta_i + hT_h,
\end{aligned}$$

which coincides with (5.11) for $a_\nu = \delta_\nu, h = h, L = L_\phi, k = 1, B = T_h$. Lemma 5.6 proves $\delta_\nu \le T_h(e^{\nu h L_\phi} - 1)/L_\phi$ because of $a_0 = \delta_0 = 0$. $\square$

So far, consistency is not assumed. From consistency we now derive convergence. Furthermore, consistency order and convergence order will coincide.

Theorem 5.19. *Let the one-step method (5.5) fulfil the Lipschitz condition (5.16), and assume consistency. Then (5.5) is also convergent:*

$$\lim_{h \to 0} \eta(x,h) = y(x).$$

If, in addition, the consistency order is p, then also the convergence is of order p.

Proof. Consistency implies $T_h \to 0$, so that convergence follows from (5.19). Consistency order p means $T_h \le Ch^p$ and implies $|\eta(x,h) - y(x)| \le \mathcal{O}(h^p)$. $\square$

---

[6] Note that perturbations in the initial value x_0 are already analysed in Theorem 5.9.

We remark that, in general, the Lipschitz condition (5.16) holds only locally. Then one argues as follows. $G := \{(x, y) : x \in [x_0, x_E], |y - y(x)| \le 1\}$ is compact. It suffices[7] to require (5.16) on G. For sufficiently small h, $T_h \frac{e^{(x-x_0)L_\phi} - 1}{L_\phi}$ in (5.19) is bounded by 1 and therefore $(x, \eta(x, h)) \in G$. A view to the proof of Lemma 5.18 shows that all intermediate arguments belong to G and therefore, (5.16) is applicable.

According to Theorem 5.19, one may be very optimistic that any consistent one-step method applied to some ordinary differential equation is working well. However, the statement concerns only the asymptotic behaviour as $h \to 0$. A problem arises if, e.g., the asymptotic behaviour is only observed for $h \le 10^{-9}$, while we want to apply the methods for a step size $h \ge 0.001$. This gives rise to stronger stability requirements (cf. §5.5.8).

5.5 Analysis of Multistep Methods

The general multistep method is defined in (5.6). Introducing formally $\alpha_r := 1$, we can rewrite the r-step method as

$$\sum_{\nu=0}^{r} \alpha_\nu \eta_{j+\nu} = h\phi(x_j, \eta_{j+r}, \eta_{j+r-1}, \ldots, \eta_j, h; f). \tag{5.20a}$$

A multistep method is called *linear* if ϕ has the particular form

$$\phi(x_j, \eta_{j+r}, \eta_{j+r-1}, \ldots, \eta_j; h; f) = \sum_{\mu=0}^{r} b_\mu \underbrace{f(x_{j+\mu}, \eta_{j+\mu})}_{=:f_{j+\mu}}. \tag{5.20b}$$

If $b_r = 0$, the linear multistep method is explicit, otherwise it is implicit.

The coefficients α_ν from (5.20a) define *the characteristic polynomial*

$$\psi(\zeta) := \sum_{\nu=0}^{r} \alpha_\nu \zeta^\nu. \tag{5.21a}$$

In the case of a linear multistep method, a further polynomial can be introduced:

$$\sigma(\zeta) := \sum_{\nu=0}^{r-1} b_\nu \zeta^\nu. \tag{5.21b}$$

Remark 5.20. *Consistency condition (5.7) is equivalent to*

$$\psi(1) = 0. \tag{5.22}$$

---

[7] Locally Lipschitz continuous functions are uniformly Lipschitz continuous on a compact set.

5.5.1 Local Discretisation Error, Consistency

Using $Y(x; \xi, \eta)$ from (5.17), we define the *local discretisation error* by

$$\tau(x, y; h) := \qquad\qquad\qquad\qquad\qquad\qquad\qquad (5.23)$$

$$\frac{1}{h}\left[\sum_{\nu=0}^{r} \alpha_\nu Y(x_{j+\nu}; x_j, y) - h\phi\Big(x_j, Y(x_{j+\nu-1}; x_j, y), \ldots, \underbrace{Y(x_j; x_j, y)}_{=y}, h; f\Big)\right].$$

Definition 5.21. *A multistep methods is called* consistent *if*

$$\sup_{x\in I} |\tau(x, y(x); h)| \to 0 \qquad (h \to 0)$$

for all $f \in C(I \times \mathbb{R})$ *with Lipschitz property (5.2). Here* $y(x)$ *is the solution of (5.1a,b). Furthermore, the multi-step method (5.20a) is called* consistent of order p *if* $|\tau(x, y(x); h)| = \mathcal{O}(h^p)$ *holds for sufficiently smooth* f.

For $f = 0$ and the initial value $y_0 = 1$, the solution is $y(x) = 1$ and, in this case, $\tau(x, y(x); h) \to 0$ simplifies to $\left(\sum_{\nu=0}^{r} \alpha_\nu - h\phi\right)/h \to 0$, implying $\sum_{\nu=0}^{r} \alpha_\nu = 0$, which is condition (5.7).

5.5.2 Convergence

Differently from the case of a one-step method, we cannot assume exact starting values $\eta_1, \ldots, \eta_{r-1}$. Therefore, we assume that all starting values are perturbed:

$$\eta_j = y(x_j) + \varepsilon_j, \qquad \varepsilon = (\varepsilon_j)_{j=0,\ldots,r-1}.$$

We denote the solution corresponding to these starting values by $\eta(x; \varepsilon, h)$.

Definition 5.22. *A multistep methods is called* convergent *if for all* $f \in C(I \times \mathbb{R})$ *with (5.2) and all starting values* y_0, *the global error satisfies*

$$\sup_{x\in I} |\eta(x; \varepsilon, h) - y(x)| \to 0 \qquad \text{for } h \to 0 \text{ and } \|\varepsilon\|_\infty \to 0.$$

A stronger requirement is that we also perturb the equations (5.20a) by ε_j for $j \geq r$:

$$\sum_{\nu=0}^{r} \alpha_\nu \eta_{j+\nu} = h\phi(x_j, \eta_{j+r-1}, \ldots, \eta_j, h; f) + h\varepsilon_{j+r} \qquad \text{for } j \geq 0. \qquad (5.24)$$

In this case, $\varepsilon = (\varepsilon_j)_{j\geq0}$ is a tuple with as many entries as grid points (note that the quantities ε_j for $j < r$ and $j \geq r$ have a quite different meaning!). Again, $\eta(x; \varepsilon, h) \to y(x)$ can be required for $h \to 0$ and $\|\varepsilon\|_\infty \to 0$.

5.5.3 Stability

We recall the characteristic polynomial ψ defined in (5.21a).

Definition 5.23. *The multistep method (5.20a) is called* stable *if all roots ζ of the characteristic polynomial ψ have the following property: either $|\zeta| < 1$ or ζ is a simple zero with $|\zeta| = 1$.*

We check three examples:
(1) The midpoint rule (5.8) yields $\psi(\zeta) = \zeta^2 - 1$. Both zeros $\zeta = \pm 1$ are simple with $|\zeta| = 1$. Therefore, the midpoint rule is stable.
(2) The two-step method (5.9) describing the extrapolation corresponds to $\psi(\zeta) = \zeta^2 - 2\zeta + 1 = (\zeta - 1)^2$. Therefore, $\zeta = 1$ is a double root and indicates instability.
(3) In the case of a one-step methods (i.e., $r = 1$), $\alpha_r = 1$ and (5.7) lead to $\psi(\zeta) = \zeta - 1$. The only zero $\zeta = 1$ satisfies the second condition in Definition 5.23. This proves the following remark corresponding to the result of Lemma 5.18.

Remark 5.24. *One-step methods are always stable in the sense of Definition 5.23.*

5.5.4 Difference Equations

The relation between the stability condition from Definition 5.23 and the multistep methods (5.20a) is not quite obvious. The connection will be given in the study of difference equations. As preparation we first discuss power bounded matrices.

5.5.4.1 Power Bounded Matrices

Definition 5.25. *Let $\|\cdot\|$ be a matrix norm. A square matrix A is power bounded[8] if*

$$\sup\{\|A^n\| : n \in \mathbb{N}\} < \infty. \tag{5.25}$$

Because of the norm equivalence in finite-dimensional vector spaces, the choice of the matrix norm $\|\cdot\|$ in Definition 5.25 is irrelevant.
To prepare the next theorem, we recall some terms from linear algebra:
Let λ be an eigenvalue of $A \in \mathbb{C}^{d \times d}$. λ has *algebraic multiplicity* $k \in \mathbb{N}_0$, if the characteristic polynomial $\det(\zeta I - A)$ contains the factor $(\zeta - \lambda)^k$, but not $(\zeta - \lambda)^{k+1}$. λ has *geometric multiplicity* $k \in \mathbb{N}_0$ if $\dim\{e \in \mathbb{C}^d : Ae = \lambda e\} = k$. The inequality 'geometric multiplicity $\leq$ algebraic multiplicity' is always valid.

---

[8] In principle, we would like to use the term 'stable matrix'; however, this notation is already used for matrices satisfying $\sup\{\|\exp(tA)\| : t > 0\} < \infty$.
  We remark that there is an extension of power boundedness to a family $\mathcal{F}$ of matrices by $\sup\{\|A^n\| : n \in \mathbb{N}, A \in \mathcal{F}\} < \infty$. Characterisations are given by the *Kreiss matrix theorem* (cf. Kreiss [13], Morton [17], and [18, Sect. 4.9]). For a generalisation, see Toh–Trefethen [22].

$\|\cdot\|$ is an *associated matrix norm* in $\mathbb{C}^{d\times d}$ if there is a vector norm $\||\cdot\||$ in $\mathbb{C}^d$ such that $\|A\| = \sup\{\|| Ax \|| / \||x\||: x \neq 0\}$ for all $A \in \mathbb{C}^{d\times d}$.

The norm $\|\cdot\|_\infty$ has two meanings. For vectors $x \in \mathbb{C}^d$, it is the maximum norm $\|x\|_\infty = \max_i |x_i|$, while for matrices it is the associated matrix norm. Because of the property $\|M\|_\infty = \max_i \sum_j |M_{ij}|$, it is also called the *row-sum (matrix) norm*.

We denote the *spectrum* of a square matrix M by

$$\sigma(M) := \{\lambda \in \mathbb{C} : \lambda \text{ eigenvalue of } M\}.$$

Exercise 5.26. *Suppose that $\|\cdot\|$ is an associated matrix norm. Prove $|\lambda| \leq \|M\|$ for all $\lambda \in \sigma(M)$.*

Theorem 5.27. *Equivalent characterisations of the power boundedness of A are (5.26a) as well as (5.26b):*

> *All eigenvalues of A satisfy either*
>   *(a) $|\lambda| < 1$ or* $\qquad\qquad\qquad\qquad\qquad\qquad\qquad\qquad$ (5.26a)
>   *(b) $|\lambda| = 1$, and λ has coinciding algebraic and geometric multiplicities.*
>
> *There is an associated matrix norm such that $\|A\| \leq 1$.* $\qquad\qquad\qquad$ (5.26b)

Proof. We shall prove (5.25) $\Rightarrow$ (5.26a) $\Rightarrow$ (5.26b) $\Rightarrow$ (5.25).

(i) Assume (5.25) and set $C_{\text{stab}} := \sup\{\|A^n\| : n \in \mathbb{N}\}$. Choose an associated matrix norm $\|\cdot\|$. Applying Exercise 5.26 to $M = A^n$, we obtain $|\lambda^n| = |\lambda|^n \leq \|A^n\| \leq C_{\text{stab}}$ for all $n \in \mathbb{N}$, hence $|\lambda| \leq 1$. Assume $|\lambda| = 1$. If λ has higher algebraic than geometric multiplicity, there is an eigenvector $e \neq 0$ and a generalised eigenvector h, so that $Ae = \lambda e$ and $Ah = e + \lambda h$. We conclude that $A^n h = \lambda^{n-1}(ne + \lambda h)$ and $\|A^n h\| = \|ne + \lambda h\| \geq n\|e\| - \|\lambda h\| \to \infty$ as $n \to \infty$, in the contradiction to $\|A^n h\| \leq C_{\text{stab}}\|h\|$. Hence, (5.26a) holds.

(ii) Assume (5.26a). Sort the eigenvalues λ_i of A such that $|\lambda_1| \leq |\lambda_2| \leq \ldots < |\lambda_{d-m+1}| = \ldots = |\lambda_d| = 1$, where $m \geq 0$ is the number the zeros with absolute value equal to one. Consider the Jordan normal form

$$J = T^{-1}AT = \begin{bmatrix} J_1 & 0 \\ 0 & D \end{bmatrix} \text{ with } \begin{cases} J_1 = \begin{bmatrix} \lambda_1 & * & & \\ & \lambda_2 & * & \\ & & \ddots & * \\ & & & \lambda_{d-m} \end{bmatrix}, \\ D = \text{diag}\{\lambda_{d-m+1}, \ldots, \lambda_d\}, \end{cases}$$

where the entries $*$ are either zero or one. Since the algebraic and geometric multiplicities of $\lambda_{d-m+1}, \ldots, \lambda_d$ coincide, D is a diagonal $m \times m$ matrix, while all eigenvalues λ_i ($i = 1, \ldots, r - m$) have absolute value < 1. Set $\Delta_\varepsilon := \text{diag}\{1, \varepsilon, \varepsilon^2, \ldots, \varepsilon^{r-1}\}$ with $\varepsilon \in (0, 1 - |\lambda_{r-m}|]$. One verifies that $\Delta_\varepsilon^{-1} J \Delta_\varepsilon$ has the row-sum norm $\|\Delta_\varepsilon^{-1} J \Delta_\varepsilon\|_\infty \leq 1$. Therefore, a transformation by $S := T\Delta_\varepsilon$ yields the norm $\|S^{-1}AS\|_\infty \leq 1$. $\|A\| := \|S^{-1}AS\|_\infty$ is the associated matrix norm corresponding to the vector norm $\|x\| := \|Sx\|_\infty$. This proves (5.26b).

(iii) Assume (5.26b). Associated matrix norms are submultiplicative; i.e., $\|A^n\| \leq \|A\|^n$, so that $\|A\| \leq 1$ implies $C_{\text{stab}} = 1 < \infty$. $\qquad\square$

So far, we have considered a single matrix. In the case of a family $\mathcal{A} \subset \mathbb{C}^{n \times n}$ of matrices, one likes to have a criterion as to whether there is a uniform bound of $\sup\{\|A^n\| : n \in \mathbb{N}, A \in \mathcal{A}\}$.

Lemma 5.28. *Let $\mathcal{A} \subset \mathbb{C}^{n \times n}$ be bounded, and suppose that there is some $\gamma < 1$ such that the eigenvalues satisfy $|\lambda_1(A)| \leq |\lambda_2(A)| \leq \ldots \leq |\lambda_n(A)| \leq 1$ and $|\lambda_{n-1}(A)| \leq \gamma < 1$ for all $A \in \mathcal{A}$. Then $\sup\{\|A^n\| : n \in \mathbb{N}, A \in \mathcal{A}\} < \infty$.*

Proof. We may choose the spectral norm $\|\cdot\| = \|\cdot\|_2$. The Schur normal form R_A is defined by $A = Q R_A Q^{-1}$, Q unitary, R_A upper triangular, with $(R_A)_{ii} = \lambda_i(A)$. By boundedness of $\mathcal{A}$, there is some $M > 0$ such that R_A is bounded entry-wise by $(R_A)_{ij} \leq R_{ij}$, where the matrix R is defined by $R_{ij} := 0$ for $i > j$, $R_{ii} := \gamma$ for

$$1 \leq i \leq n-1, \ R_{nn} := 1, \ R_{ij} := M \text{ for } i < j; \text{i.e., } R = \begin{bmatrix} \gamma & \ldots & M & M \\ 0 & \ddots & M & M \\ 0 & \ldots & \gamma & M \\ 0 & \ldots & 0 & 1 \end{bmatrix}. \text{ It is easy to}$$

verify that $\|A^n\|_2 = \|R_A^n\|_2 \leq \|R^n\|_2$. Since R is power bounded, we have proved a uniform bound. $\quad\square$

5.5.4.2 Solution Space $\mathcal{F}_0$

The set $\mathcal{F} = \mathbb{C}^{\mathbb{N}_0}$ consists of sequences $\mathbf{x} = (x_j)_{j \in \mathbb{N}_0}$ of complex numbers. We are looking for sequences $\mathbf{x} \in \mathcal{F}$ satisfying the following difference equation:

$$\sum_{\nu=0}^{r} \alpha_\nu x_{j+\nu} = 0 \qquad \text{for all } j \geq 0, \text{where } \alpha_r = 1. \tag{5.27}$$

Lemma 5.29. *(a) $\mathcal{F}$ forms a linear vector space.*
(b) $\mathcal{F}_0 := \{\mathbf{x} \in \mathcal{F} \text{ satisfies (5.27)}\}$ is a linear subspace of $\mathcal{F}$ with $\dim(\mathcal{F}_0) = r$.

Proof. (i) The vector space properties of $\mathcal{F}$ and $\mathcal{F}_0$ are trivial. It remains to prove $\dim \mathcal{F}_0 = r$. We define $\mathbf{x}^{(i)} \in \mathcal{F}$ for $i = 0, 1, \ldots, r-1$ by the initial values $x_j^{(i)} = \delta_{ij}$ for $j \in \{0, \ldots, r-1\}$. For $j \geq r$ we use (5.27) to define

$$x_j^{(i)} := \sum_{\nu=0}^{r-1} \alpha_\nu x_{j-r+\nu}^{(i)} \qquad \text{for all } j \geq r. \tag{5.28}$$

Obviously, $\mathbf{x}^{(i)}$ satisfies (5.27); i.e., $\mathbf{x}^{(i)} \in \mathcal{F}_0$ for $0 \leq i \leq r-1$.

(ii) Assume $\sum_i \beta_i \mathbf{x}^{(i)} = 0$. Evaluating the entries at $j \in \{0, \ldots, r-1\}$, we obtain $0 = \sum_i \beta_i x_j^{(i)} = \sum_i \beta_i \delta_{ij} = \beta_j$, which proves linear independence.

(iii) For each $\mathbf{x} \in \mathcal{F}_0$ we define $\mathbf{y} := \mathbf{x} - \sum_{i=0}^{r-1} x_i \mathbf{x}^{(i)} \in \mathcal{F}_0$, which by definition satisfies $y_0 = y_1 = \ldots = y_{r-1} = 0$. Analogous to (5.28), these initial values lead to $y_j = 0$ for all $j \geq r$. The result $\mathbf{y} = 0$ proves that $\{\mathbf{x}^{(0)}, \ldots, \mathbf{x}^{(r-1)}\}$ already spans $\mathcal{F}_0$. Together, $\dim(\mathcal{F}_0) = r$ follows. $\quad\square$

5.5.4.3 Representation of the Solutions

Remark 5.30. *(a) Let $\zeta_0 \in \mathbb{C}$ be a root of the polynomial $\psi(\zeta) = \sum_{\nu=0}^{r} \alpha_\nu \zeta^\nu$.*
Then $\mathbf{x} = (\zeta_0^j)_{j \in \mathbb{N}_0}$ is a solution of (5.27); i.e., $\mathbf{x} \in \mathcal{F}_0$.
(b) Suppose that ψ has r different zeros $\zeta_i \in \mathbb{C}$, $i = 1, \ldots, r$. Then the solutions
$\mathbf{x}^{(i)} = (\zeta_i^j)_{j \in \mathbb{N}_0}$ form a basis of the space $\mathcal{F}_0$.

Proof. (i) Inserting $x_j = \zeta_0^j$ into (5.27) yields

$$\sum_{\nu=0}^{r} \alpha_\nu x_{j+\nu} = \sum_{\nu=0}^{r} \alpha_\nu \zeta_0^{j+\nu} = \zeta_0^j \sum_{\nu=0}^{r} \alpha_\nu \zeta_0^\nu = \zeta_0^j \psi(\zeta_0) = 0;$$

i.e., $\mathbf{x} \in \mathcal{F}_0$.
   (ii) By Part (i), $\mathbf{x}^{(i)} \in \mathcal{F}_0$ holds. One easily verifies that the solutions $\mathbf{x}^{(i)}$ are linearly independent. Because of $\dim \mathcal{F}_0 = r$, $\{\mathbf{x}^{(i)} : 1 \le i \le r\}$ forms a basis. $\square$

In the case of Remark 5.30, the zeros are simple. It remains to discuss the case of multiple zeros.
   We recall that the polynomial ψ has an (at least) k-fold zero ζ_0 if and only if $\psi(\zeta_0) = \psi'(\zeta_0) = \ldots = \psi^{(k-1)}(\zeta_0) = 0$. The Leibniz rule yields

$$\left(\frac{\mathrm{d}}{\mathrm{d}\zeta}\right)^\ell \left(\zeta^j \psi(\zeta)\right) = 0 \quad \text{at} \quad \zeta = \zeta_0 \qquad \text{for } 0 \le \ell \le k-1.$$

The explicit representation of $\left(\frac{\mathrm{d}}{\mathrm{d}\zeta}\right)^\ell \left(\zeta^j \psi(\zeta)\right)$ reads

$$0 = \left(\zeta^j \psi(\zeta)\right)^{(\ell)}\Big|_{\zeta=\zeta_0} = \sum_{\nu=0}^{r} \alpha_\nu \zeta_0^{j+\nu-\ell} (j+\nu)(j+\nu-1) \cdot \ldots \cdot (j+\nu-\ell+1).$$

$$(5.29)$$

   Define $\mathbf{x}^{(\ell)}$ for $\ell \in \{0, 1, \ldots, k-1\}$ via $x_j^{(\ell)} = \zeta_0^j j (j-1) \cdot \ldots \cdot (j-\ell+1)$.
Insertion into the difference equation (5.27) yields

$$\sum_{\nu=0}^{r} \alpha_\nu x_{j+\nu}^{(\ell)} = \sum_{\nu=0}^{r} \alpha_\nu \zeta_0^{j+\nu} (j+\nu)(j+\nu-1) \cdot \ldots \cdot (j+\nu-\ell+1).$$

This is ζ_0^ℓ times the expression in (5.29); hence $\sum_{\nu=0}^{r} \alpha_\nu x_{j+\nu}^{(\ell)} = 0$; i.e., $\mathbf{x}^{(\ell)} \in \mathcal{F}_0$.

Remark 5.31. *Let $\zeta_0 \ne 0$ be a zero of ψ with multiplicity k.*
(a) Then $\mathbf{x}^{(0)}, \ldots, \mathbf{x}^{(k-1)}$ with $x_j^{(\ell)} = \zeta_0^j \prod_{\nu=0}^{\ell-1}(j-\nu)$ ($\ell = 0, \ldots, k-1$) are k linearly independent solutions of (5.27).
(b) Similarly, $\hat{x}_j^{(\ell)} = \zeta_0^j j^\ell$ represent k linearly independent solutions of (5.27).

Proof. (i) $\mathbf{x}^{(\ell)} \in \mathcal{F}_0$ is shown above. The linear independence for $\ell = 0, \ldots, k-1$
follows, e.g., from the different growth of $x_j^{(\ell)}/\zeta_0^j$ for $j \to \infty$.

(ii) $\left\{\prod_{\nu=0}^{\ell-1}(x-\nu) : \ell = 0, \ldots, k-1\right\}$ as well as $\left\{x^\ell : \ell = 0, \ldots, k-1\right\}$ are bases of the polynomials of degree $\leq k-1$. Therefore $\{\mathbf{x}^{(0)}, \ldots, \mathbf{x}^{(k-1)}\}$ and $\{\hat{\mathbf{x}}^{(0)}, \ldots, \hat{\mathbf{x}}^{(k-1)}\}$ span the same space. $\square$

The case $\zeta_0 = 0$ is excluded in Remark 5.31, since the previous definition leads to $x_j^{(\ell)} = 0$ for $j \geq \min\{1-\ell, 0\}$ and therefore does not yield linearly independent solutions.

Remark 5.32. *Let $\zeta_0 = 0$ be a k-fold zero of ψ. Then $\mathbf{x}^{(i)}$ with $x_j^{(i)} = (\delta_{ij})_{j\in\mathbb{N}_0}$ $(i = 0, \ldots, k-1)$ are k linearly independent solutions of (5.27).*

Proof. Use $\psi(0) = \psi'(0) = \ldots = \psi^{(k-1)}(0) = 0$ to obtain $\alpha_0 = \ldots = \alpha_{k-1} = 0$. This shows that $\sum_{\nu=0}^r \alpha_\nu x_{j+\nu}^{(i)} = \sum_{\nu=k}^r \alpha_\nu x_{j+\nu}^{(i)}$. By definition, $x_{j+\nu}^{(i)} = 0$ holds for $\nu \geq k$; i.e., (5.27) is satisfied. Obviously, the $\mathbf{x}^{(i)}$ are linearly independent. $\square$

Theorem 5.33. *Let ζ_i $(i = 1, \ldots, m)$ be the different zeros of the polynomial $\psi(\zeta) = \sum_{\nu=0}^r \alpha_\nu \zeta^\nu$ with the corresponding multiplicities k_i. Any ζ_i gives rise to k_i solutions of (5.27), which are defined in Remark 5.31, if $\zeta_i \neq 0$, and in Remark 5.32, if $\zeta_i = 0$. Altogether, r linearly independent solutions spanning $\mathcal{F}_0$ are characterised.*

Proof. The linear independence of the constructed solutions is left as an exercise. In Theorem 5.33, $\sum k_i = r$ solutions are described. They form a basis of $\mathcal{F}_0$, since, according to Lemma 5.29, $\dim \mathcal{F}_0 = r$ holds. $\square$

5.5.4.4 Stability

Definition 5.34. *The difference equation (5.27) is called* stable *if any solution of (5.27) is bounded with respect to the supremum norm:*

$$\|\mathbf{x}\|_\infty := \sup_{j\in\mathbb{N}_0} |x_j| < \infty \qquad \text{for all } \mathbf{x} \in \mathcal{F}_0.$$

Remark 5.35. *A characterisation equivalent to Definition 5.34 reads as follows: there is a constant C, so that*

$$\|\mathbf{x}\|_\infty \leq C \max_{j=0,\ldots,r-1} |x_j| \qquad \text{for all } \mathbf{x} \in \mathcal{F}_0. \tag{5.30}$$

Proof. (i) Eq. (5.30) implies $\|\mathbf{x}\|_\infty < \infty$, since $\max_{j=0,\ldots,r-1} |x_j|$ is always finite.

(ii) We choose the basis $\mathbf{x}^{(i)} \in \mathcal{F}_0$ as in (ii) of the proof of Lemma 5.29. $\mathbf{x}$ has a representation $\mathbf{x} = \sum_{i=0}^{r-1} x_i \mathbf{x}^{(i)}$. Assuming stability in the sense of Definition 5.34, we have $C_i := \|\mathbf{x}^{(i)}\|_\infty < \infty$ and therefore also $C := \sum_{i=0}^{r-1} C_i < \infty$. The estimate $\|\mathbf{x}\|_\infty = \|\sum_{i=0}^{r-1} x_i \mathbf{x}^{(i)}\|_\infty \leq \sum_{i=0}^{r-1} |x_i| \|\mathbf{x}^{(i)}\|_\infty = \sum_{i=0}^{r-1} C_i |x_i| \leq C \max_{j=0,\ldots,r-1} |x_j|$ proves (5.30). $\square$

Obviously, all solutions of (5.27) are bounded if and only if all basis solutions given in Theorem 5.33 are bounded. The following complete list of disjoint cases refers to the zeros ζ_i of ψ and their multiplicities k_i.

1. $|\zeta_i| < 1$: all sequences $(\zeta_i^j j^\ell)_{j \in \mathbb{N}_0}$ with $0 \le \ell < k_i$ are zero sequences and therefore bounded.
2. $|\zeta_i| > 1$: for all sequences lead to $\lim |\zeta_i^j j^\ell| = \infty$; i.e., they are unbounded.
3. $|\zeta_i| = 1$ and $k_i = 1$ (simple zero): $(\zeta_i^j)_{j \in \mathbb{N}_0}$ is bounded by 1 in absolute value.
4. $|\zeta_i| = 1$ and $k_i > 1$ (multiple zero): $\lim |\zeta_i^j j^\ell| = \infty$ holds for $1 \le \ell \le k_i - 1$; i.e., the sequences are unbounded.

Therefore, the first and third cases characterise the stable situations, while the second and fourth cases lead to instability. This proves the next theorem.

Theorem 5.36. *The difference equation (5.27) is stable if and only if ψ satisfies the stability condition from Definition 5.23.*   •

5.5.4.5 Companion Matrix

Definition 5.37. *The companion matrix of the polynomial $\psi(\zeta) = \sum_{\nu=0}^{r} \alpha_\nu \zeta^\nu$ is of size $r \times r$ and is equal to*

$$
A = \begin{bmatrix}
0 & 1 & & \\
& \ddots & \ddots & \\
& & 0 & 1 \\
-\alpha_0 & \cdots & -\alpha_{r-2} & -\alpha_{r-1}
\end{bmatrix}. \tag{5.31}
$$

Remark 5.38. *(a)* $\det(\zeta I - A) = \psi(\zeta)$.
(b) Any solution $(x_j)_{j \in \mathbb{N}_0}$ of the difference equation (5.27) satisfies (5.32) and, vice versa, (5.32) implies (5.27):

$$
\begin{bmatrix}
x_{j+1} \\
x_{j+2} \\
\vdots \\
x_{j+r}
\end{bmatrix}
= A
\begin{bmatrix}
x_j \\
x_{j+1} \\
\vdots \\
x_{j+r-1}
\end{bmatrix}. \tag{5.32}
$$

By (5.32) we can formally reformulate the r-step method as a one-step method for the r-tuple $X_j := (x_j, \ldots, x_{j+r-1})^\top$. Therefore, the stability behaviour must be expressed by the properties of A. An obvious connection is described in the next remark.

Remark 5.39. *The difference equation (5.27) is stable if and only if A is a power bounded matrix.*

Proof. Let $\|\cdot\|$ denote a vector norm as well as the associated matrix norm. The tuple X_j satisfies $X_j = A X_{j-1}$ (cf. (5.32)). In particular, $X_n = A^n X_0$ holds.

Uniform boundedness of $|x_j|$ and $\|X_j\|$ are equivalent. If (5.27) is stable, $\|A^n X_0\|$ is uniformly bounded for all X_0 with $\|X_0\| \leq 1$ and all $n \in \mathbb{N}$; i.e., A is a power bounded matrix. On the other hand, $C_{\text{stab}} := \sup\{\|A^n\| : n \in \mathbb{N}\} < \infty$ yields the estimate $\|X_n\| \leq C_{\text{stab}} \|X_0\|$ and therefore stability. $\quad\square$

Using (5.26b), we obtain the next statement.

Lemma 5.40. *The difference equation (5.27) is stable if and only if there is an associated matrix norm, so that $\|A\| \leq 1$ holds for the companion matrix A from (5.31).*

5.5.4.6 Estimates of Inhomogeneous Solutions

Theorem 5.41. *Suppose that the difference equation is stable and that the initial values fulfil $|x_j| \leq \alpha$ for $0 \leq j^{\circ} \leq r - 1$. If the sequence $(x_j)_{j \in \mathbb{N}_0}$ satisfies the inhomogeneous difference equation $\sum_{\nu=0}^{r} \alpha_\nu x_{j+\nu} = \beta_{j+r}$ with*

$$|\beta_{j+r}| \leq \beta + \gamma \max\{|x_\mu| : 0 \leq \mu \leq j + r - 1\}$$

for some $\gamma \geq 0$, then there exist k and k' such that

$$|x_j| \leq k k' \alpha e^{jk\gamma} + \begin{cases} jk\beta & \text{for } \gamma = 0, \\ \frac{\beta}{\gamma}\left(e^{j\gamma k} - 1\right) & \text{for } \gamma > 0. \end{cases}$$

Proof. (i) Let A be the companion matrix. $\|\cdot\|$ denotes the vector norm in $\mathbb{R}^r$ as well as the associated matrix norm $\|\cdot\|$ from Lemma 5.40. Because of the norm equivalence,

$$\|X\|_\infty \leq k \|X\|, \quad \|X\| \leq k' \|X\|_\infty \qquad \text{for all } X \in \mathbb{R}^r$$

holds for suitable k, k'.

(ii) Set $X_j := (x_j, \ldots, x_{j+r-1})^\top$ and $e = (0, \ldots, 0, 1)^\top \in \mathbb{R}^r$. According to Lemma 5.40, $\|A\| \leq 1$ holds. Since a scaling of the vector norm does not change the associated matrix norm, we assume without loss of generality that $\|e\| = 1$. The difference equation $\sum_{\nu=0}^{r} \alpha_\nu x_{j+\nu} = \beta_{j+r}$ is equivalent to

$$X_{j+1} = AX_j + \beta_{j+r} e.$$

(iii) Set $\xi_j := \max\{|x_\mu| : 0 \leq \mu \leq j\}$. By definition of X_μ, also $\xi_j = \max_{0 \leq \mu \leq j-r+1} \|X_\mu\|_\infty$ is valid for $j \geq r - 1$. It follows that

$$\|X_{j+1}\| = \|AX_j + \beta_{j+r} e\| \leq \underbrace{\|A\|}_{\leq 1} \|X_j\| + |\beta_{j+r}| \underbrace{\|e\|}_{=1} \leq \|X_j\| + |\beta_{j+r}|$$

$$\leq \|X_j\| + \beta + \gamma \xi_{j+r-1}.$$

Define η_j by

$$\eta_0 := \|X_0\|, \qquad \eta_{j+1} := \eta_j + \beta + \gamma \xi_{j+r-1}. \tag{5.33}$$

Obviously, $\|X_j\| \leq \eta_j$ and $\eta_{j+1} \geq \eta_j$ $(j \geq 0)$ hold. The estimate

$$\xi_{j+r-1} = \max_{0 \leq \mu \leq j+r-1} |x_\mu| = \max_{0 \leq \mu \leq j} \|X_\mu\|_\infty$$

$$\leq \max_{0 \leq \mu \leq j} k \|X_\mu\| \leq k \max_{0 \leq \mu \leq j} \eta_\mu \underset{\eta_{j+1} \geq \eta_j}{=} k\eta_j$$

together with the definition (5.33) of η_{j+1} yields

$$\eta_{j+1} \leq (1 + \gamma k)\, \eta_j + \beta.$$

Apply Lemma 5.6 to this inequality. The corresponding quantities in (5.11) are $\nu \equiv j, a_\nu \equiv \eta_j, h \equiv 1, L \equiv k\gamma, B \equiv \beta$. Lemma 5.6 yields the inequality

$$\eta_j \leq \eta_0 e^{jk\gamma} + \left\{ \begin{matrix} j\beta & \text{if } \gamma = 0 \\ \frac{\beta}{k\gamma} \left(e^{jk\gamma} - 1 \right) & \text{if } \gamma > 0 \end{matrix} \right\} \qquad (j \in \mathbb{N}_0).$$

Furthermore, $\eta_0 = \|X_0\| \leq k' \|X_0\|_\infty \leq k'\alpha$ and $|x_j| \leq \|X_j\|_\infty \leq k \|X_j\| \leq k\eta_j$ holds. Together, the assertion of the theorem follows. □

5.5.5 Stability and Convergence Theorems

We shall show that convergence and stability of multistep methods are almost equivalent. For exact statements one needs a further assumption concerning the connection of $\phi(x_j, \eta_{j+r-1}, \ldots, \eta_j, h; f)$ and f. A very weak assumption is

$$f = 0 \quad \Longrightarrow \quad \phi(x_j, \eta_{j+r-1}, \ldots, \eta_j, h; f) = 0. \tag{5.34}$$

This assumption is satisfied, in particular, for the important class of *linear r-step methods*:

$$\phi(x_j, \eta_{j+r}, \ldots, \eta_j, h; f) = \sum_{\mu=0}^{r} b_\mu f_{j+\mu} \qquad \text{with } f_k = f(x_k, \eta_k). \tag{5.35}$$

Theorem 5.42 (stability theorem). *Suppose (5.34). Then the convergence from Definition 5.22 implies stability.*

Proof. (i) We choose $f = 0$ and the starting value $y_0 = 0$. Therefore, $y = 0$ is the exact solution of the initial-value problem, while the discrete solution satisfies the equations $\sum_{\nu=0}^{r} a_\nu \eta_{j+\nu} = 0$ with initial values $\eta_0 = \varepsilon_0, \ldots, \eta_{r-1} = \varepsilon_{r-1}$ (cf. Definition 5.22). Since $\sum_{\nu=0}^{r} a_\nu \eta_{j+\nu} = 0$ is the difference equation (5.27), $(\eta_j)_{j \in \mathbb{N}_0} \in \mathcal{F}_0$ holds. However, we have to note that the multistep method visits only the finite section $(\eta_j)_{0 \leq j \leq J(h)}$ with $J(h) = \lfloor (x_E - x_0)/h \rfloor$, since those j satisfy $x_j \in I$.

(ii) For the indirect proof, assume instability. Then an unbounded solution $\mathbf{x} \in \mathcal{F}_0$ exists. The divergence

$$C(h) := \max\{|x_j| : 0 \le j \le J(h)\} \to \infty \quad \text{for } h \to 0$$

follows from $J(h) \to \infty$ for $h \to 0$ and $\|\mathbf{x}\|_\infty = \infty$. Choose the initial perturbation $\varepsilon = (\varepsilon_j)_{j=0,\ldots,r-1} := (x_j/C(h))_{j=0,\ldots,r-1}$. Obviously, $\|\varepsilon\|_\infty \to 0$ holds for $h \to 0$. For this initial perturbation the multistep method produces the solution $(\eta_j)_{0 \le j \le J(h)}$ with $\eta_j = \frac{1}{C(h)} x_j$. Since

$$\sup_{x \in I} |\eta(x; \varepsilon, h) - y(x)| = \frac{1}{C(h)} \max\{|x_j| : 0 \le j \le J(h)\} = 1,$$

this error does not tend to 0 as $h \to 0$, $\|\varepsilon\|_\infty \to 0$, in contradiction to the assumed convergence.   □

For the reverse direction we need a Lipschitz condition corresponding to (5.16) in the case of a one-step method:

for each $f \in C(I \times \mathbb{R})$ with (5.2) there is $L_\phi \in \mathbb{R}$ such that   (5.36)

$$|\phi(x_j, u_{r-1}, \ldots, u_0, h; f) - \phi(x_j, v_{r-1}, \ldots, v_0, h; f)| \le L_\phi \max_{i=0,\ldots,r-1} |u_i - v_i|.$$

Remark 5.43. *Condition (5.36) is satisfied for linear r-step methods (5.35).*

Theorem 5.44 (convergence theorem). *Let (5.36) be valid. Furthermore, the multistep method is supposed to be consistent and stable. Then it is convergent (even in the stronger sense as discussed below Definition 5.22).*

Proof. The initial error is defined by $\eta_j = y(x_j) + \varepsilon_j$ for $j = 0, \ldots, r-1$. The multistep formula with additional errors $h\varepsilon_{j+r}$ is described in (5.24). The norm of the error is $\|\varepsilon\|_\infty := \max\{|\varepsilon_j| : 0 \le j \le J(h)\}$ with $J(h)$ as in the previous proof (the usual convergence leads to $\varepsilon_j = 0$ for $r \le j \le J(h)$, only in the case described in brackets can $\varepsilon_j \ne 0$ appear for $r \le j \le J(h)$). We have to show that $\eta(x; \varepsilon, h) \to y(x)$ for $h \to 0$, $\|\varepsilon\|_\infty \to 0$.

The error is denoted by $e_j := \eta_j - y(x_j)$. The initial values for $0 \le j \le r-1$ are $e_j = \varepsilon_j$. The equation

$$\sum_{\nu=0}^{r} \alpha_\nu y(x_{j+\nu}) - h\phi(x_j, y(x_{j+r-1}), \ldots, y(x_j), h; f) = h\tau(x_j, y(x_j); h) =: h\tau_{j+r}$$

containing the local discretisation error τ_{j+r} is equivalent to (5.23). We form the difference between the latter equation and (5.24) for $j \ge r$ and obtain

$$\sum_{\nu=0}^{r} \alpha_\nu e_{j+\nu} = \beta_{j+r} := h\left(\varepsilon_{j+r} - \tau_{j+r}\right)$$

$$+ h\left[\phi(x_j, \eta_{j+r-1}, \ldots, \eta_j, h; f) - \phi(x_j, y(x_{j+r-1}), \ldots, y(x_j), h; f)\right].$$

From (5.36) we infer that

$$|\beta_{j+r}| \leq hL_\phi \max_{j \leq \mu \leq, j+r-1} |e_\mu| + h\left(\|\varepsilon\|_\infty + \|\tau\|_\infty\right),$$

where $\tau = (\tau_j)_{r \leq j \leq J(h)}$, $\|\tau\|_\infty = \max_{r \leq j \leq J(h)} |\tau_j|$.

The consistency condition $\sup_{x \in I} \tau(x, y(x); h) \to 0$ implies $\|\tau\|_\infty \to 0$ for $h \to 0$.
The assumptions of Theorem 5.41 hold with

$$x_j = e_j, \quad \alpha = \|\varepsilon\|_\infty, \quad \beta = h\left(\|\varepsilon\|_\infty + \|\tau\|_\infty\right), \quad \gamma = hL_\phi,$$

so that the theorem yields

$$|e_j| \leq kk' \|\varepsilon\|_\infty e^{jhL_\phi k} + \frac{\|\varepsilon\|_\infty + \|\tau\|_\infty}{L_\phi} \left(e^{jhL_\phi k} - 1\right) \qquad (5.37)$$

in the case of $hL_\phi > 0$ (the case $hL_\phi = 0$ is analogous). The product jh in the exponent is to be interpreted as $x_j - x_0$ and therefore bounded by $x_E - x_0$ (or it is constant and equal to x in the limit process $j = n \to \infty$, $h := (x - x_0)/n$). As part of the definition of convergence, $\|\tau\|_\infty \to 0$ (consistency) and $\|\varepsilon\|_\infty \to 0$ holds for $h \to 0$. According to (5.37), e_j converges uniformly to zero; i.e., $\sup_{x \in I} |\eta(x; \varepsilon, h) - y(x)| \to 0$. $\square$

Corollary 5.45. *In addition to the assumptions in Theorem 5.44 assume consistency of order p. Then also the convergence order is p, provided that the initial errors are sufficiently small:*

$$|\eta(x; \varepsilon, h) - y(x)| \leq C\left(h^p + \max_{j=0}^{r-1} |\varepsilon_j|\right).$$

Proof. We have $\|\varepsilon\|_\infty = \max_{0 \leq j \leq r-1} |\varepsilon_j|$ and $\|\tau\|_\infty \leq \mathcal{O}(h^p)$. Inequality (5.37) proves the desired error bound. $\square$

5.5.6 Construction of Optimal Multistep Methods

5.5.6.1 Examples

The *Adams–Bashforth methods* are explicit linear r-step methods of the form

$$\eta_{j+r} = \eta_{j+r-1} + h \sum_{\mu=0}^{r-1} b_\mu f_{j+\mu} \qquad (5.38)$$

(cf. (5.20b)). The associated characteristic polynomial is

$$\psi(\zeta) = \zeta^r - \zeta^{r-1} = \zeta^{r-1}(\zeta - 1).$$

Its zeros are $\zeta_1 = \ldots = \zeta_{r-1} = 0$, $\zeta_r = 1$, so that the Adams–Bashforth methods are stable. The coefficients b_μ ($\mu = 0, \ldots, r-1$) can be used to make the local consistency error as small as possible. The optimal choice yields a multistep method[9] of order r.

Exercise 5.46. (a) Euler's method is the optimal Adams–Bashforth method for $r = 1$.
(b) What are the optimal coefficients b_0, b_1 in (5.38) for $r = 2$?

The general explicit linear two-step method

$$\eta_{j+2} = -\alpha_1 \eta_{j+1} - \alpha_0 \eta_j + h \left[b_1 f_{j+1} + b_0 f_j \right]$$

contains four free parameter. Because of the side condition (5.7) (i.e., $\alpha_0 + \alpha_1 = 1$), there remain three degrees of freedom. Thus, the optimal choice $\alpha_0 = -5$, $\alpha_0 = 2$, $\alpha_1 = 4$, $b_1 = 4$ of the coefficients can reach consistency of order $p = 3$. The resulting method is

$$\eta_{j+2} = -4\eta_{j+1} + 5\eta_j + h \left[4 f_{j+1} + 2 f_j \right]. \tag{5.39}$$

j	x_j	$\eta_j - y(x_j)$
2	0.02	-0.16_{10}-8
3	0.03	$+0.50_{10}$-8
4	0.04	-0.30_{10}-7
5	0.05	$+0.14_{10}$-6
$\vdots$	$\vdots$	$\vdots$
99	0.99	$+0.13_{10}$+60
100	1.00	-0.65_{10}+60

The associated polynomial is

$$\psi(\zeta) = \zeta^2 + 4\zeta - 5 = (\zeta - 1)(\zeta + 5).$$

The root -5 proves instability. We demonstrate that this instability is clearly observed in practice. We apply (5.39) to the initial-value problem $y' = -y$, $y(0) = 1 =: \eta_0$ ($\Rightarrow y(x) = e^{-x}$) and choose the exact value $\eta_1 := e^{-h}$ to avoid any further error. The step size in $I = [0, 1]$ is chosen by $h = 0.01$. The root $\zeta = -5$ is responsible for alternating signs of the error and for the explosive increase ($5^{98} = 3.2_{10}68$).

5.5.6.2 Stable Multistep Methods of Optimal Order

The previous example shows that one cannot use all coefficients α_ν, b_μ from (5.6) and (5.35) in order to maximise the consistency order. Instead stability is a side condition, when we optimise α_ν, b_μ. The characterisation of optimal stable multistep methods is due to Dahlquist [2].

Theorem 5.47. (a) If $r \geq 1$ is odd, the highest consistency order of a stable linear r-step method is $p = r + 1$.
(b) If $r \geq 2$ is even, the highest consistency order is $p = r + 2$. In this case, all roots of the characteristic polynomial ψ have absolute value 1.

---

[9] Note the following advantage of multistep methods compared with one-step methods: In spite of the increased consistency order r, only one function value $f_{j+r-1} = f(x_{j+r-1}, \eta_{j+r-1})$ needs to be evaluated per grid point x_j; the others are known from the previous steps.

5.5.6.3 Proof

Step 1: A stable linear r-step method of order $p = r + 1$ is existing.

For this purpose, substitute the variable ζ of the polynomials $\psi(\zeta)$ and $\sigma(\zeta)$ from (5.21a,b) by z:

$$\zeta = \frac{1+z}{1-z}, \qquad z = \frac{\zeta-1}{\zeta+1}. \tag{5.40a}$$

This defines the functions

$$p(z) := \left(\frac{1-z}{2}\right)^r \psi\left(\frac{1+z}{1-z}\right), \qquad s(z) = \left(\frac{1-z}{2}\right)^r \sigma\left(\frac{1+z}{1-z}\right). \tag{5.40b}$$

Remark 5.48. *(a) $p(z)$ and $s(z)$ are polynomials of degree $\le r$.*
(b) If $\zeta_0 \neq -1$ is a root of $\psi(\zeta)$ with multiplicity k, then $p(z)$ has the root $z_0 = \frac{\zeta_0-1}{\zeta_0+1}$ again of multiplicity k.
(c) The transformation (5.40a) maps the complex unit circle $|\zeta| < 1$ onto the left half-plane $\Re e\, z < 0$; in particular, $\zeta = 1$ is mapped onto $z = 0$, and $\zeta = -1$ onto $z = \infty$.

Proof. (i) Part (c) can be verified directly. For Part (a) note that $p(z)$ is a linear combination of the polynomials $(\frac{1+z}{1-z})^\nu (1-z)^r = (1-z)^{r-\nu}(1+z)^\nu$ for $0 \le \nu \le r$, which are all of degree r. Similarly for s.

(ii) Let k be the multiplicity of ζ_0. Then $\psi(\zeta) = (\zeta - \zeta_0)^k \psi_0(\zeta)$ holds with some $\psi_0(\zeta_0) \neq 0$. Hence,

$$p(z) = \left(\frac{1-z}{2}\right)^r \left(\frac{1+z}{1-z} - \zeta_0\right)^k \psi_0\left(\frac{1+z}{1-z}\right)$$

$$= (1 + z - (1-z)\zeta_0)^k \left[\frac{1}{2^r}(1-z)^{r-k} \psi_0\left(\frac{1+z}{1-z}\right)\right]$$

$$= (z - z_0)^k \left[\frac{(1-z)^{r-k}}{2^r(\zeta_0+1)^k} \psi_0\left(\frac{1+z}{1-z}\right)\right].$$

Since $\zeta_0 = \infty$ is excluded, $z_0 = 1$ is not a root and the bracket $[\ldots]$ does not vanish at $z = z_0$. This proves Part (b). $\square$

Because of stability and $\psi(1) = 0$ (cf. (5.22)), $\zeta = 1$ is a simple root. By Remark 5.48c, $p(z)$ has a simple root at $z = 0$. Hence, p is of the form

$$p(z) = \alpha_1 z + \alpha_2 z^2 + \ldots + \alpha_\ell z^\ell \qquad \text{with } \alpha_1 \neq 0, \ell = \text{degree}(p) \le k. \tag{5.41a}$$

Without loss of generality, we may assume that

$$\alpha_1 > 0. \tag{5.41b}$$

(otherwise scale the equation of the multistep method by -1 changing $\psi(\zeta)$ into $-\psi(\zeta)$). We shall prove that

$$\alpha_\mu \geq 0 \qquad \text{for all } 1 \leq \mu \leq r. \tag{5.41c}$$

For this purpose, we denote the zeros of p by $z_\nu = x_\nu + i\,y_\nu$ ($x_\nu, y_\nu \in \mathbb{R}$). Then,

$$p(z) = \alpha_\ell z \prod_\nu (z - z_\nu) = \alpha_\ell z \prod_{\nu \text{ with } y_\nu = 0} (z - x_\nu) \prod_{\nu \text{ with } y_\nu \neq 0} \left((z - x_\nu)^2 + y_\nu^2 \right)$$
$$\tag{5.41d}$$

must hold, where the last product is taken over all pairs of conjugate complex zeros. Stability implies $|\zeta_\nu| \leq 1$, which by Remark 5.48c implies $\Re\, z_\nu \leq 0$; i.e., $x_\nu \leq 0$. Because of $z - x_\nu = z + |x_\nu|$, all polynomial coefficients of $p(z)$ must carry the same sign (or vanish). Hence, (5.41b) implies (5.41c).

We define the holomorphic function

$$\varphi(\zeta) := \frac{\psi(\zeta)}{\log \zeta} - \sigma(\zeta). \tag{5.42}$$

Note that because of $\psi(1) = 0$, φ has no singularity at $\zeta = 1$. A related function is

$$g(z) := \left(\frac{1-z}{2} \right)^r \varphi\left(\frac{1+z}{1-z} \right) = \frac{p(z)}{\log \frac{1+z}{1-z}} - s(z). \tag{5.43}$$

Theorem 5.49. *Assume $\psi(1) = 0$ (cf. (5.22)). Then the linear multistep method (5.20b) has the (local) consistency order p if and only if $\zeta = 1$ is a p-fold root of φ.*

Proof. Assume consistency order p. We choose the differential equation $y' = y$. Then we have (up to a factor) $z(t) = e^t$ and

$$\tau(x, y, h) = \frac{\sum_{\nu=0}^{r} a_\nu e^{x+\nu h} - h \sum_{\nu=0}^{r} b_\nu e^{x+\nu h}}{h} = e^x \frac{\sum_{\nu=0}^{r} a_\nu (e^h)^\nu - h \sum_{\nu=0}^{r} b_\nu (e^h)^\nu}{h}$$

$$= e^x \left(\frac{\psi(e^h)}{h} - \sigma(e^h) \right) = e^x \varphi(e^h).$$

Set $\delta := e^h - 1 = h \cdot e^{\theta h}$ ($0 < \theta < h$ from the mean value theorem). Hence, δ can be estimated from both sides by const $\cdot\, h$. The Taylor expansion of $\varphi(e^h) = \varphi(1 + \delta)$ around $\delta = 0$ exists, since φ is holomorphic at $\zeta = 1$:

$$\varphi(e^h) = \varphi(1) + \varphi'(1)\delta + \ldots + \frac{\varphi^{(p-1)}(1)}{(p-1)!} \delta^{p-1} + \mathcal{O}(\delta^p).$$

Since $\delta^k \sim h^k$, we conclude from $\tau(x, y, h) = \mathcal{O}(h^p) = \mathcal{O}(\delta^p)$ that terms involving δ^k with $k < p$ cannot appear; i.e., $\varphi(1) = \varphi'(1) = \ldots = \varphi^{(p-1)}(1) = 0$. Hence, 1 is a p-fold zero of φ.

On the other hand, if 1 is a p-fold zero, the Taylor expansion shows that $\tau(x, y, h) = e^x \varphi(e^h) = \mathcal{O}(\delta^p) = \mathcal{O}(h^p)$ and therefore the method has consistency order p. $\square$

Concerning g, the following implications from (5.43) are valid:

$$g(z) \text{ has a } p\text{-fold zero at } z = 0$$
$$\Leftrightarrow \quad \varphi(\zeta) \text{ has a } p\text{-fold zero at } \zeta = 1$$
$$\Leftrightarrow \quad p = \text{consistency order.}$$

We recall the function $p(z)$ from (5.40b). Since $p(z)/\log \frac{1+z}{1-z}$ is holomorphic at $z = 0$, there is a power series

$$\frac{z}{\log \frac{1+z}{1-z}} \frac{p(z)}{z} = \beta_0 + \beta_1 z + \beta_2 z^2 + \ldots \qquad (5.44a)$$

The function g connects p and s. To obtain a p-fold zero of g at $z = 0$, s must be of the form $s(z) = \sum_{\mu=0}^r \beta'_\mu z^\mu$ with $\beta'_\mu = \beta_\mu$ for $0 \le \mu \le p-1$. Since we consider the case $p = r+1 > r = \text{degree}(s)$ of Theorem 5.47a, the polynomial $s(z)$ is already uniquely determined:

$$s(z) = \beta_0 + \beta_1 z + \ldots + \beta_r z^r, \qquad (5.44b)$$

where $r = p - 1$. Polynomial s from (5.44b) determines a unique σ (cf. (5.40b)). This ends the construction of a stable method of order $p = r + 1$.

Step 2: For odd r, there is no stable method of consistency order $p > r + 1$.

Let the method have an order $p > r + 1$. The comparison of (5.44a) and (5.44b) shows that in (5.44a)

$$\beta_\mu = 0 \qquad \text{for } r + 1 \le \mu \le p - 1 \qquad (5.44c)$$

must hold (the first β_μ for $0 \le \mu \le r$ can be made to zero by the choice (5.44b)). The function $z/\log \frac{1+z}{1-z}$ is an even function in z, so that the power series becomes

$$\frac{z}{\log \frac{1+z}{1-z}} = c_0 + c_2 z^2 + c_4 z^4 + \ldots$$

Let α_μ be the coefficients from (5.41a), where we set $\alpha_\mu := 0$ for $\mu > \ell = \text{degree}(p)$. Comparing the coefficients of both sides in (5.44a), we obtain

$$\beta_0 = c_0 \alpha_1, \qquad \ldots \qquad \beta_{2\nu} = c_0 \alpha_{2\nu+1} + c_2 \alpha_{2\nu-1} + \ldots + c_{2\nu} \alpha_1,$$
$$\beta_1 = c_0 \alpha_2, \qquad \ldots \qquad \beta_{2\nu+1} = c_0 \alpha_{2\nu+2} + c_2 \alpha_{2\nu} + \ldots + c_{2\nu} \alpha_2.$$

We shall prove that $c_{2\nu} < 0$ for all $\nu \ge 1$. Hence, for odd r it follows that

$$\beta_{r+1} = \underbrace{c_0 \alpha_{r+2}}_{=0} + \underbrace{c_2}_{<0} \underbrace{\alpha_r}_{\ge 0} + \underbrace{c_4}_{<0} \underbrace{\alpha_{r-2}}_{\ge 0} + \ldots + \underbrace{c_{r+1}}_{<0} \underbrace{\alpha_1}_{>0} < 0 \qquad (5.44d)$$

in contradiction to $\beta_{r+1} = 0$ (cf. (5.44c)).

For the proof of $c_{2\nu} < 0$ we need the following lemma of Kaluza [12].

Lemma 5.50. *Let* $f(t) = \sum_{\nu=0}^{\infty} A_\nu t^\nu$ *and* $g(t) = \sum_{\nu=0}^{\infty} B_\nu t^\nu$ *be power series with the properties*

$$f(t)g(t) \equiv 1, \quad A_\nu > 0 \ (\nu \geq 0), \quad A_{\nu+1}A_{\nu-1} > A_\nu^2 \ (\nu \geq 1).$$

Then $B_\nu < 0$ *holds for all* $\nu \geq 1$.

Proof. (i) The assumption $A_{\nu+1}A_{\nu-1} > A_\nu^2$ can be written as $\frac{A_{\nu+1}}{A_\nu} > \frac{A_\nu}{A_{\nu-1}}$ for $\nu \geq 1$. From this we conclude that $\frac{A_{n+1}}{A_n} > \frac{A_{n-\nu+1}}{A_{n-\nu}}$ for $1 \leq \nu \leq n$. The latter inequality can be rewritten as

$$A_{n+1}A_{n-\nu} - A_n A_{n-\nu+1} > 0. \tag{5.45}$$

(ii) Without loss of generality assume $A_0 = 1$. This implies $B_0 = 1$. Comparison of the coefficients in $f(t)g(t) \equiv 1$ proves that

$$0 = A_n + \sum_{\nu=1}^{n} B_\nu A_{n-\nu} \ (n \geq 1), \quad -B_{n+1} = A_{n+1} + \sum_{\nu=1}^{n} B_\nu A_{n-\nu+1} \ (n \geq 0).$$

For $n = 0$, the latter identity shows that $B_1 < 0$. Multiply the first equation by A_{n+1}, the second by $-A_n$, and add: $A_n B_{n+1} = \sum_{\nu=1}^{n} B_\nu (A_{n+1}A_{n-\nu} - A_n A_{n-\nu+1})$. Thanks to (5.45), $B_{n+1} < 0$ follows by induction. □

We apply this lemma to $f(z^2) = \frac{1}{z} \log \frac{1+z}{1-z} = 2 + \frac{2}{3}z^2 + \frac{2}{5}z^4 + \ldots$ The coefficients $A_\nu = \frac{2}{2\nu+1} > 0$ satisfy

$$A_{\nu+1}A_{\nu-1} = \frac{2}{2\nu+3} \cdot \frac{2}{2\nu-1} = \frac{4}{(2\nu+1)^2 - 4} > \frac{4}{(2\nu+1)^2} = A_\nu^2$$

and therefore the supposition of the lemma. Since $B_\nu = c_{2\nu}$, the assertion of Step 2 is proved.

Step 3: For even r, stable methods are characterised as in Theorem 5.47b.

For even r, the sum corresponding to (5.44d) becomes

$$\beta_{r+1} = \underbrace{c_0 \alpha_{r+2}}_{=0} + \underbrace{c_2}_{<0} \underbrace{\alpha_r}_{\geq 0} + \underbrace{c_4}_{<0} \underbrace{\alpha_{r-2}}_{\geq 0} + \ldots + \underbrace{c_r}_{<0} \underbrace{\alpha_2}_{\geq 0} \leq 0,$$

where $\beta_{r+1} = 0$ holds if and only if $\alpha_2 = \alpha_4 = \ldots = \alpha_r = 0$. The latter property is equivalent to $p(z)$ being odd:

$$p(z) = -p(-z) \qquad \text{for all } z.$$

Hence, each root z_ν of p corresponds to a root $-z_\nu$. As stability requires $\Re e\, z_\nu \leq 0$, we obtain $\Re e\, z_\nu = 0$ for all z_ν, corresponding to the condition $|\zeta_\nu| = 1$. On the other hand, we have the reverse statement: If $|\zeta_\nu| = 1$ for all root of ψ, it follows that $\Re e\, z_\nu = 0$ and

$$p(z) = \text{const} \cdot z \prod_{\nu} \left(z^2 - (\Im m \, z_\nu)^2 \right)$$

and therefore $p(z) = -p(-z)$. Hence, the order $p = r + 2$ can be obtained.

Order $p = r + 3$ (r even) cannot be reached, since by Step 2 $\beta_{r+2} < 0$.

5.5.7 Further Remarks

5.5.7.1 Systems of Differential Equations

The previous considerations concerning stability remain unchanged for systems $y' = f(x, y)$, where y and f take values in $\mathbb{R}^n$. Nevertheless, further complications can appear.

First, we consider the case of *stiff systems*. The previous convergence analysis provides statements for the limit $h \to 0$. The proven error estimates hold for h *sufficiently small*. A more practical question is: how large can we choose h, when does the asymptotic regime start? For instance, the differential equation $y' = -y$ has the positive and decreasing solution e^{-x}, while the explicit Euler scheme (5.4) with step size $h = 3$ yields the values $\eta_j = (-2)^j$, which are oscillatory in sign and increasing. A simple analysis shows that for $h < 1$ also the discrete solution is positive and decreasing. Now assume the system $y' = Ay$ of linear differential equations with $A = \text{diag}\{-1, -1000\} \in \mathbb{R}^{2 \times 2}$. The dominant part of the solution is $\binom{y_1}{0} e^{-x}$ related to the eigenvalue one, since the other component $\binom{0}{y_2} e^{-1000x}$ decays very strongly. As in the one-dimensional example from above, one would like to use a step size of the size $h < 1$. However, the second component enforces the inequality $h < 1/1000$. This effect appears for general linear systems, if A possesses one eigenvalue of moderate size and another one with strongly negative real part, or for nonlinear systems $y' = f(x, y)$, where $A(x) := \partial f / \partial y$ has similar properties. This leads to the definition of *A-stability* (or absolute stability; cf. [20], [9], [3]). Good candidates for A-stable methods are implicit ones (cf. 5.4.1). Implicit methods have to solve a linear system of the form $Az = b$. Here, another problem arises. For instance, the spatial discretisation of a parabolic initial value problem (see Chap. 6) yields a large stiff system of ordinary differential equations. The solution of the large linear system requires further numerical techniques (cf. [6]).

Next, we consider the formulation $By' = Cy$ of a linear system or, in the general nonlinear case, $F(x, y, y') = 0$. If B is regular (or F solvable with respect to y'), we regain the previous system with $A := B^{-1}C$. If, however, $B \in \mathbb{R}^{n \times n}$ is singular with rank k, the system $By' = Cy$ consists of a mixture of $n - k$ differential equations and k algebraic side conditions. In this case, the system is called a differential-algebraic equations (DAE; [14]). In between there are singularly perturbed systems, where $B = B_0 + \varepsilon B_1$ is regular for $\varepsilon > 0$, but ε is small and B_0 is singular.

5.5.8 Other Stability Concepts

The analysis of numerical schemes for ordinary differential equations has created many further variants of stability definitions. Besides this, there are stability conditions (e.g., Lyapunov stability) which are not connected with discretisations, but with the (undiscretised) differential equation and its dynamical behaviour (cf. [8, §I.13], [11, §X], [7], [21]).

References

1. Butcher, J.C.: A history of Runge-Kutta methods. Appl. Numer. Math. **20**, 247–260 (1996)
2. Dahlquist, G.: Convergence and stability in the numerical integration of ordinary differential equations. Math. Scand. **4**, 33–53 (1956)
3. Deuflhard, P., Bornemann, F.: Scientific computing with ordinary differential equations. Springer, New York (2002)
4. Deuflhard, P., Bornemann, F.: Numerische Mathematik II. Gewöhnliche Differentialgleichungen, 3rd ed. Walter de Gruyter, Berlin (2008)
5. Gautschi, W.: Numerical Analysis. An Introduction. Birkhäuser, Boston (1997)
6. Hackbusch, W.: Iterative solution of large sparse systems of equations. Springer, New York (1994)
7. Hahn, W.: Stability of Motion. Springer, Berlin (1967)
8. Hairer, E., Nørsett, S.P., Wanner, G.: Solving Ordinary Differential Equations I, 2nd ed. North-Holland, Amsterdam (1993)
9. Hairer, E., Wanner, G.: Solving Ordinary Differential Equations II. Springer, Berlin (1991)
10. Heun, K.: Neue Methoden zur approximativen Integration der Differentialgleichungen einer unabhängigen Veränderlichen. Z. für Math. und Phys. **45**, 23–38 (1900)
11. Heuser, H.: Gewöhnliche Differentialgleichungen. Teubner, Stuttgart (1991)
12. Kaluza, T.: Über Koeffizienten reziproker Potenzreihen. Math. Z. **28**, 161–170 (1928)
13. Kreiss, H.O.: Über die Stabilitätsdefinition für Differenzengleichungen die partielle Differentialgleichungen approximieren. BIT **2**, 153–181 (1962)
14. Kunkel, P., Mehrmann, V.: Differential-algebraic equations - analysis and numerical solution. EMS, Zürich (2006)
15. Kutta, W.M.: Beitrag zur näherungsweisen Integration totaler Differentialgleichungen. Z. für Math. und Phys. **46**, 435–453 (1901)
16. Lipschitz, R.O.S.: Lehrbuch der Analysis, Vol. 2. Cohen, Bonn (1880)
17. Morton, K.W.: On a matrix theorem due to H. O. Kreiss. Comm. Pure Appl. Math. **17**, 375–379 (1964)
18. Richtmyer, R.D., Morton, K.W.: Difference Methods for Initial-value Problems, 2nd ed. John Wiley & Sons, New York (1967). Reprint by Krieger Publ. Comp., Malabar, Florida, 1994
19. Runge, C.D.T.: Ueber die numerische Auflösung von Differentialgleichungen. Math. Ann. **46**, 167–178 (1895)
20. Stetter, H.J.: Analysis of Discretization Methods for Ordinary Differential Equations. North-Holland, Amsterdam (1973)
21. Terrel, W.J.: Stability and Stabilization - An Introduction. Princeton University Press, Princeton and Oxford (2009)
22. Toh, K.C., Trefethen, L.N.: The Kreiss matrix theorem on a general complex domain. SIAM J. Matrix Anal. Appl. **21**, 145–165 (1999)
23. Werner, H., Arndt, H.: Gewöhnliche Differentialgleichungen. Eine Einführung in Theorie und Praxis. Springer, Berlin (1986)

Chapter 6
Instationary Partial Differential Equations

The analysis presented in this chapter evolved soon after 1950, when discretisations of hyperbolic and parabolic differential equations had to be developed. Most of the material can be found in Richtmyer–Morton [21], see also Lax–Richtmyer [16]. All results concern linear differential equations. In the case of hyperbolic equations there is a crucial difference between linear and nonlinear problems, since in the nonlinear case many unpleasant features may occur that are unknown in the linear case. Concerning general hyperbolic conservation laws, we refer, e.g., to Kröner [14] and LeVeque [19]. However, even in the nonlinear case, the linearised problems should satisfy the stability conditions described here.

6.1 Introduction and Examples

6.1.1 Notation, Problem Setting, Function Spaces

Replacing the scalar ordinary differential equation by a *linear system* of ordinary differential equations, one obtains $y' = Ay + f$, where—in the simplest case—A is a constant $N \times N$ matrix, and y and f have values in $\mathbb{R}^N$ (or $\mathbb{C}^N$). In this case, the previous results concerning one-step and multistep methods can easily be generalised. The situation changes if the (bounded) matrix A is replaced by an unbounded differential operator, as we do now.

The independent variable x of the ordinary differential equation is renamed by t ('time' variable). The differential operator A contains differentiations with respect to spatial variables $x_1, \ldots, x_d$. Although $d = 3$ is the realistic case, we restrict our considerations mainly to $d = 1$. The case $d > 1$ will be discussed in §6.5.6.1.

Notation 6.1. *The desired solution is denoted by u (instead of y). The independent variables are t and x. The classical notation for u depending on t and x is $u(t, x)$. Let B be a space of functions in the variable x. Then $u(t)$ denotes the function $u(t, \cdot) \in B$ (partially evaluated at t). Therefore, $u(t, x)$ and $u(t)(x)$ are equivalent*

W. Hackbusch, *The Concept of Stability in Numerical Mathematics*,
Springer Series in Computational Mathematics 45, DOI 10.1007/978-3-642-39386-0_6,
© Springer-Verlag Berlin Heidelberg 2014

notations. Let $I = [0, T]$ be the time interval in which t varies. If $D_A \subset B$ is the domain[1] of the differential operators A, the partial differential equation takes the following form: Find a continuous function $u : I \to D_A \subset B$ such that

$$\frac{\partial}{\partial t} u(t) = A u(t) \qquad \text{for all } t \in I. \tag{6.1a}$$

The initial-value condition is given by

$$u(0) = u_0 \qquad \text{for some } u_0 \in D_A \subset B. \tag{6.1b}$$

Concerning the differential operator A, we discuss two model cases:

$$A := a \frac{\partial}{\partial x} \quad (a \neq 0) \qquad \text{and} \qquad A := a \frac{\partial^2}{\partial x^2} \quad (a > 0). \tag{6.2}$$

In what follows, the domain of $u(\cdot, \cdot)$ is the set

$$\Sigma = I \times \mathbb{R} \qquad \text{with } I = [0, T]. \tag{6.3}$$

Here the time t varies in $I = [0, T]$, while the spatial variable x varies in $\mathbb{R}$. I corresponds to the interval $I = [x_0, x_E]$ from §5.1. The spatial domain $\mathbb{R}$ is chosen as the unbounded domain to avoid boundary conditions.[2]

We restrict our considerations to two Banach spaces, generally denoted by B:

- $B = C(\mathbb{R})$, space of the complex-valued, uniformly[3] continuous functions with finite supremum norm $\|v\|_B = \|v\|_\infty = \sup\{|v(x)| : x \in \mathbb{R}\}$.
- $B = L^2(\mathbb{R})$, space of the complex-valued, measurable, and square-integrable functions. This means that the L^2 norm

$$\|v\|_B = \|v\|_2 = \sqrt{\int_\mathbb{R} |v(x)|^2 \, \mathrm{d}x}$$

is finite. This Banach space is also a Hilbert space with the scalar product

$$(u, v) := \int_\mathbb{R} u(x) \overline{v(x)} \mathrm{d}x.$$

For both cases of (6.2) we shall show that the initial-value problem is solvable.

---

[1] The *domain* of a differential operator A is $D_A = \{v \in B : Av \in B \text{ is defined}\}$. Often, it suffices to choose a smaller, dense set $B_0 \subset D_A$ and to extend the results continuously onto D_A.

[2] A similar situation arises if the solutions are assumed to be 2π-periodic in x. This corresponds to the bounded domain $\Sigma = I \times [0, 2\pi]$ with *periodic boundary condition* $u(t, 0) = u(t, 2\pi)$. In the 2π-periodic case, the spaces are

$$C_{\text{per}}(\mathbb{R}) := \{v \in C(\mathbb{R}) : v(x) = v(x + 2\pi) \text{ for all } x \in \mathbb{R}\},$$
$$L^2_{\text{per}}(\mathbb{R}) := \{v \in L^2(\mathbb{R}) : v(x) = v(x + 2\pi) \text{ for almost all } x \in \mathbb{R}\}.$$

[3] Note that limits of uniformly continuous functions are again uniformly continuous.

6.1.2 The Hyperbolic Case $A = a\partial/\partial x$

First we choose $B = C(\mathbb{R})$. The domain of $A = a\frac{\partial}{\partial x}$ is the subspace $B_0 = C^1(\mathbb{R})$, which is dense in B. The partial differential equation[4]

$$\frac{\partial}{\partial t}u = a\frac{\partial}{\partial x}u$$

is of hyperbolic type.[5] The solution of problem (6.1a,b) can be described directly.

Lemma 6.2. *For any $u_0 \in B_0 = C^1(\mathbb{R})$, the unique solution of the initial-value problem (6.1a,b) is given by $u(t, x) := u_0(x + at)$.*

Proof. (i) By $\frac{\partial}{\partial t}u = \frac{\partial}{\partial t}u_0(x+at) = au_0'(x+at)$ and $a\frac{\partial}{\partial x}u_0(x+at) = au_0'(x+at)$, the differential equation (6.1a) is satisfied. The initial value is $u(0, x) := u_0(x)$.

(ii) Concerning uniqueness, we transform to the (characteristic) direction:

$$\xi = x + at, \quad \tau = t, \quad U(\tau, \xi) := u(t(\tau, \xi), x(\tau, \xi))$$

with the inverse transformation $t(\tau, \xi) = \tau$, $x(\tau, \xi) = \xi - a\tau$. The chain rule yields $\frac{\partial}{\partial \tau}U = \frac{\partial}{\partial t}u\frac{\partial}{\partial \tau}t + \frac{\partial}{\partial x}u\frac{\partial}{\partial \tau}x = \frac{\partial}{\partial t}u - a\frac{\partial}{\partial x}u = 0$. With respect to the new variables, Eq. (6.1a) becomes the ordinary differential equation $\frac{\partial}{\partial \tau}U(\tau, \xi) = 0$ (for each $\xi \in \mathbb{R}$). Hence, there is a unique constant solution $U(\tau, \xi) = U(0, \xi) = u(t(0, \xi), x(0, \xi)) = u(0, \xi) = u_0(\xi)$. $\square$

In the case of the space $B = L^2(\mathbb{R})$, we choose $B_0 := C_0^\infty(\mathbb{R}) \subset D_A$ (or $B_0 := C^1(\mathbb{R}) \cap L^2(\mathbb{R})$) as a dense subspace in B. Here $C_0^\infty(\mathbb{R})$ is the set of all infinitely often differentiable functions with compact support (cf. Footnote 13 on page 56). Lemma 6.2 holds also for this B_0.

Remark 6.3. *Let $t \geq 0$ be arbitrary. If the initial value u_0 belongs to $C^1(\mathbb{R})$ or $C_0^\infty(\mathbb{R})$, then also the solution $u(t)$ belongs to $C^1(\mathbb{R})$ or $C_0^\infty(\mathbb{R})$, respectively. Furthermore,*

$$\|u(t)\|_B = \|u_0\|_B$$

holds for $\|\cdot\|_B = \|\cdot\|_\infty$ as well as for $\|\cdot\|_B = \|\cdot\|_2$.

Proof. Since $u(t)$ is a shifted version of u_0 and a shift does not change the norm $\|\cdot\|_B$, the assertions follow. $\square$

Exercise 6.4. *Extend the previous statement to any L^p norm*

$$\|u\|_p = \sqrt[p]{\int_\mathbb{R} |u(x)|^p \, dx} \quad (1 \leq p < \infty).$$

---

[4] The case of $a = 0$ is exceptional, since then $u_t = 0$ can be considered as a family of ordinary differential equations for each $x \in \mathbb{R}$. Hence, the theory of §5 applies again.

[5] Concerning the definition of types of partial differential equations, see Hackbusch [8, §1].

6.1.3 The Parabolic Case $A = \partial^2/\partial x^2$

The parabolic differential equation $\frac{\partial}{\partial t} u = a \frac{\partial^2 u}{\partial x^2}$ is called the *heat equation*, since it describes the evolution of the temperature u as a function of time t and space x in the case an infinite wire (one-dimensional case!). The factor $a > 0$ is the thermal conductivity. If a is constant, we may assume $a = 1$ (otherwise, transform by $t \mapsto at$ or $x \mapsto \sqrt{a}x$):

$$\frac{\partial u}{\partial t} = \frac{\partial^2 u}{\partial x^2} \qquad \text{for } t > 0. \tag{6.4}$$

Lemma 6.5. *The solution of (6.4) with a continuous initial value (6.1b) is given by*

$$u(t, x) = \frac{1}{\sqrt{4\pi t}} \int_{-\infty}^{\infty} u_0(\xi) \exp\left(\frac{-(x-\xi)^2}{4t}\right) d\xi \qquad \text{for } t > 0 \text{ and } x \in \mathbb{R}. \tag{6.5}$$

Proof. (i) One verifies that $\frac{1}{\sqrt{4\pi t}} \exp\left(\frac{-(x-\xi)^2}{4t}\right)$ is a solution of (6.4) for each $\xi \in \mathbb{R}$ and $t > 0$. Since the integrand in (6.5) decays exponentially, one can interchange integration and differentiation, and obtains that u from (6.5) satisfies (6.4).

(ii) It remains to prove that $\lim_{t \searrow 0} u(t, x) = u_0(x)$. Note that

$$\frac{1}{\sqrt{4\pi t}} \int_{-\infty}^{\infty} \exp\left(\frac{-\zeta^2}{4t}\right) d\zeta = 1 \qquad \text{for } t > 0 \tag{6.6}$$

(cf. [27, p. 187]). Substitution $\zeta = \xi - x$ yields

$$u(t, x) = \frac{1}{\sqrt{4\pi t}} \int_{-\infty}^{\infty} u_0(\xi) \exp\left(\frac{-(x-\xi)^2}{4t}\right) d\xi$$

$$= u_0(x) + \frac{1}{\sqrt{4\pi t}} \int_{-\infty}^{\infty} [u_0(\xi) - u_0(x)] \exp\left(\frac{-(x-\xi)^2}{4t}\right) d\xi.$$

If we can prove that the last term tends to zero as $t \searrow 0$, the desired statement $\lim_{t \searrow 0} u(t, x) = u_0(x)$ follows.

Let x and $\varepsilon > 0$ be fixed. Because of continuity of u_0, there is a $\delta > 0$ such that $|u_0(\xi) - u_0(x)| \leq \varepsilon/2$ for all $|\xi - x| \leq \delta$. We split the integral into the sum of three terms:

$$I_1(t, x) := \frac{1}{\sqrt{4\pi t}} \int_{x-\delta}^{x+\delta} [u_0(\xi) - u_0(x)] \exp\left(\frac{-(x-\xi)^2}{4t}\right) d\xi,$$

$$I_2(t, x) := \frac{1}{\sqrt{4\pi t}} \int_{-\infty}^{x-\delta} [u_0(\xi) - u_0(x)] \exp\left(\frac{-(x-\xi)^2}{4t}\right) d\xi,$$

$$I_3(t, x) := \frac{1}{\sqrt{4\pi t}} \int_{x+\delta}^{\infty} [u_0(\xi) - u_0(x)] \exp\left(\frac{-(x-\xi)^2}{4t}\right) d\xi.$$

The first integral is bounded by

$$|I_1(t,x)| \leq \frac{1}{\sqrt{4\pi t}} \int_{x-\delta}^{x+\delta} |u_0(\xi) - u_0(x)| \exp\left(\frac{-(x-\xi)^2}{4t}\right) d\xi$$

$$\leq \frac{\varepsilon/2}{\sqrt{4\pi t}} \int_{x-\delta}^{x+\delta} \exp\left(\frac{-(x-\xi)^2}{4t}\right) d\xi$$

$$\leq \frac{\varepsilon/2}{\sqrt{4\pi t}} \int_{-\infty}^{\infty} \exp\left(\frac{-(x-\xi)^2}{4t}\right) d\xi \underset{(6.6)}{=} \frac{\varepsilon}{2}.$$

Set $C := \sup_{x \in \mathbb{R}} |u_0(x)| < \infty$. Then I_2 is bounded by

$$|I_2(t,x)| \leq \frac{1}{\sqrt{4\pi t}} \int_{-\infty}^{x-\delta} |u_0(\xi) - u_0(x)| \exp\left(\frac{-(x-\xi)^2}{4t}\right) d\xi$$

$$\leq \frac{2C}{\sqrt{4\pi t}} \int_{-\infty}^{x-\delta} \exp\left(\frac{-(x-\xi)^2}{4t}\right) d\xi$$

$$\underset{\tau=(x-\xi)/\sqrt{4t}}{=} \frac{2C}{\sqrt{\pi}} \int_{\delta/\sqrt{4t}}^{\infty} \exp\left(-\tau^2\right) d\tau.$$

Since the improper integral $\int_{-\infty}^{\infty} \exp\left(-\tau^2\right) d\tau$ exists, $\int_{R}^{\infty} \exp\left(-\tau^2\right) d\tau$ tends to zero as $R \to \infty$. Therefore, a sufficiently small $t > 0$ yields $|I_2(t,x)| \leq \frac{\varepsilon}{4}$. We obtain the same bound $|I_3(t,x)| \leq \frac{\varepsilon}{4}$ for I_3. Together,

$$\left| \frac{1}{\sqrt{4\pi t}} \int_{-\infty}^{\infty} [u_0(\xi) - u_0(x)] \exp\left(\frac{-(x-\xi)^2}{4t}\right) d\xi \right| \leq \frac{\varepsilon}{2} + \frac{\varepsilon}{4} + \frac{\varepsilon}{4} = \varepsilon$$

holds for sufficiently small $t > 0$. As x and ε are arbitrarily chosen,

$$\lim_{t \searrow 0} \frac{1}{\sqrt{4\pi t}} \int_{-\infty}^{\infty} [u_0(\xi) - u_0(x)] \exp\left(\frac{-(x-\xi)^2}{4t}\right) d\xi = 0$$

is proved for all x. $\square$

The representation (6.5) shows that the solution $u(t)$ at $t > 0$ is infinitely often differentiable, although the initial value u_0 is only continuous. However, the solution exists only for $t > 0$, not for $t < 0$. Note the different property in the hyperbolic case, where the representation of the solution from Lemma 6.2 holds for all $t \in \mathbb{R}$.

In the hyperbolic case, the norm $\|u(t)\|_B$ is independent of t, whereas in the parabolic case, only a monotonicity statement holds.

Lemma 6.6. *Let $u(t) \in B = C(\mathbb{R})$ be a solution of (6.4). Then the inequality $\|u(t)\|_\infty \leq \|u(0)\|_\infty$ holds for all $t \geq 0$.*

Proof. Set $C := \|u(0)\|_\infty$ and let $t > 0$. From (6.5) we infer that

$$|u(t,x)| \leq \frac{1}{\sqrt{4\pi t}} \int_{-\infty}^{\infty} |u_0(\xi)| \exp\left(\frac{-(x-\xi)^2}{4t}\right) d\xi$$

$$\leq \frac{C}{\sqrt{4\pi t}} \int_{-\infty}^{\infty} \exp\left(\frac{-(x-\xi)^2}{4t}\right) d\xi \underset{(6.6)}{=} C;$$

hence, $\|u(t)\|_\infty \leq C$. $\square$

Lemma 6.7. *Let $u(t) \in B = L^2(\mathbb{R})$ be a solution of (6.4). Then $\|u(t)\|_2 \leq \|u(0)\|_2$ holds for all $t \geq 0$.*

Proof. It is sufficient to restrict to $u(0) \in D_A$. One concludes either from (6.5) or from general considerations (cf. (6.7e)) that $\frac{\partial^2 u}{\partial x^2} \in L^2(\mathbb{R})$ for $t \geq 0$, so that the following integrals exist. Let $t'' \geq t' \geq 0$. Because of

$$\int_{\mathbb{R}} u(t'', x)^2 dx - \int_{\mathbb{R}} u(t', x)^2 dx = \int_{\mathbb{R}} \int_{t'}^{t''} \frac{\partial}{\partial t}\left(u(t', x)^2\right) dt dx$$

$$= 2 \int_{\mathbb{R}} \int_{t'}^{t''} u(t, x) \frac{\partial u(t, x)}{\partial t} dt dx = 2 \int_{\mathbb{R}} \int_{t'}^{t''} u(t, x) \frac{\partial^2 u(t, x)}{\partial x^2} dt dx$$

$$= 2 \int_{t'}^{t''} \int_{\mathbb{R}} u(t, x) \frac{\partial^2 u(t, x)}{\partial x^2} dx dt = -2 \int_{t'}^{t''} \int_{\mathbb{R}} \left(\frac{\partial u(t, x)}{\partial x}\right)^2 dx dt \leq 0,$$

$\|u(t)\|_2^2$ is weakly decreasing. $\square$

6.2 Semigroup of Solution Operators

In the following, $B_0 = C^\infty(\mathbb{R}) \cap B$ is chosen as a dense subset of B. We know that $u_t = Au$ has a classical solution for any initial value $u_0 \in B_0$. A solution is called classical or strong if $u_t(t)$ and $Au(t)$ are elements of B for all $t \geq 0$. We recall that $\|u(t)\|_B \leq \|u_0\|_B$ has been proved.

Obviously, the mapping from $u_0 \in B_0$ into the solution $u(t)$ at some fixed $t \geq 0$ is linear. This defines the *solution operator*

$$T(t) : u_0 \mapsto u(t) \qquad \text{for } t \geq 0. \tag{6.7a}$$

So far, $T(t)$ is defined on B_0. From $\|u(t)\|_B \leq \|u_0\|_B$ and the density of B_0 we infer that $T(t)$ is a bounded operator and can be extended uniquely and continuously onto B.

Remark 6.8. *Suppose that $T(t) \in \mathcal{L}(B_0, B)$ and that B_0 is dense in B. Then $T(t)$ can be extended uniquely and continuously onto B. The extended $T(t) \in \mathcal{L}(B, B)$ and the original mapping $T(t) \in \mathcal{L}(B_0, B)$ have equal operator norms; i.e., $\sup\{\|T(t)v\|_B / \|v\|_B : 0 \neq v \in B_0\} = \sup\{\|T(t)v\|_B / \|v\|_B : 0 \neq v \in B\}$.*

Proof. Let $u_0 \in B$. There is a sequence $v_{0,n} \in B_0$ with $\lim_{n\to\infty} \|u_0 - v_{0,n}\|_B = 0$. One verifies that $v_n := T(t)v_{0,n}$ forms a Cauchy sequence, and therefore it defines a unique $u := \lim_{n\to\infty} v_n$. The definition $T(t)u_0 =: u$ defines the continuous extension $T(t) \in \mathcal{L}(B, B)$ of the desired form. $\square$

Because of

$$T(t) \in \mathcal{L}(B, B), \tag{6.7b}$$

any initial value $u_0 \in B$ leads to a function $u(t) := T(t)u_0 \in B$. If $u_0 \in B \backslash B_0$, the resulting function is called a 'generalised' or 'weak' solution in contrast to the strong solution mentioned above. Note that the descriptions $u(t, x) := u_0(x + at)$ from Lemma 6.2 and of $u(t, x)$ by (6.5) make sense for any $u_0 \in B$.

Next, we show the semigroup property

$$T(t)T(s) = T(t + s) \qquad \text{for all } t, s \geq 0. \tag{6.7c}$$

For this purpose consider the strong solution $u(\tau) = T(\tau)u_0$ for an initial value $u_0 \in B_0$ and fix some $s \geq 0$. Then $\hat{u}_0 := u(s)$ equals $T(s)u_0$. Set $\hat{u}(t) := u(t+s)$. Since $\hat{u}_t(t) = u_t(t + s) = Au(t + s) = A\hat{u}(t)$ and $\hat{u}(0) = u(s) = \hat{u}_0$, we conclude that $\hat{u}(t) = T(t)\hat{u}_0$; i.e., $T(t + s)u_0 = u(t + s) = \hat{u}(t) = T(t)\hat{u}_0 = T(t)T(s)u_0$ holds for all $u_0 \in B_0$. Since B_0 is dense in B, the identity (6.7c) follows.

The operators A and $T(t)$ commute:

$$AT(t) = T(t)A \qquad \text{on } D_A \tag{6.7d}$$

(see Footnote 1 for the domain D_A of A). For a proof, consider the strong solution $u(t) = T(t)u_0$ for any $u_0 \in B_0$. The third line in

$$AT(t)u_0 = Au(t) = u_t(t)$$
$$= \lim_{h \searrow 0} \frac{u(t + h) - u(t)}{h} = \lim_{h \searrow 0} \frac{1}{h} \left[T(t + h) - T(t) \right] u_0$$
$$= \lim_{h \searrow 0} T(t) \left[T(h) - I \right] u_0 = T(t) \left[\lim_{h \searrow 0} \frac{u(h) - u(0)}{h} \right]$$
$$= T(t)u_t(0) = T(t)Au(0) = T(t)Au_0$$

uses the continuity of $T(t)$ (cf. (6.7b)). Since $T(t)Au_0 \in B$ is defined for all $u_0 \in D_A$, also $AT(t)$ has this property. This proves[6]

$$T(t) : D_A \mapsto D_A \qquad \text{for } t \geq 0. \tag{6.7e}$$

The definition of $T(t)$ yields

$$T(0) = I \qquad \text{(identity)} \tag{6.7f}$$

for $t = 0$; i.e., $\{T(t) : t \geq 0\}$ is a semigroup with neutral element. It is called the semigroup generated by A. The generating operator A can be regained via $Av = \lim_{t \searrow 0} \left[(T(t)v - v) / t \right]$ for all $v \in D_A$. Another notation for $T(t)$ is $e^{tA} = \exp(tA)$.

---

[6] The semigroups of hyperbolic and parabolic problems have different properties. In the parabolic case, $T(t) : D_A \mapsto B$ holds for positive t (even $T(t) : D_A \to C^\infty(\mathbb{R}) \cap B$) as can be seen from (6.5), whereas in the hyperbolic case the smoothness does not improve with increasing t.

Finally, we discuss the behaviour of generalised solutions $u(t) = T(t)u_0$ with $u_0 \in B$ for $t \to 0$. In Lemma 6.5 u_0 is assumed to be continuous (not necessarily uniformly continuous) and point-wise convergence $u(t, x) \to u_0(x)$ for all $x \in \mathbb{R}$ is shown. Assuming uniform continuity, we obtain uniform convergence by the same proof; i.e., $u(t) = T(t)u_0 \to u_0$ in B as $t \to 0$. Under the resulting property $\lim \|[T(t) - T(0)] u_0\|_B \to 0$ for all $u_0 \in B$ the semigroup is called continuous. The same situation happens in the hyperbolic case for any $u_0 \in C(\mathbb{R})$.

Above we used the inequality $\|u(t)\|_B \leq \|u_0\|_B$ for $t \geq 0$. The constant one can be replaced by any bound and the range $t \geq 0$ may be reduced to any interval $[0, \tau]$, $\tau > 0$.

Exercise 6.9. *Suppose that $K_\tau := \sup_{0 \leq t \leq \tau} \|T(t)\|_{B \leftarrow B} < \infty$ for some $\tau > 0$. Prove $\|T(t)\|_{B \leftarrow B} \leq K_\tau^{\lceil t/\tau \rceil}$ for any $t \geq 0$, where $\lceil x \rceil := \min\{n \in \mathbb{Z} : x \leq n\}$.*

In the next subsections, we shall refer to the following inequality (6.8), which holds with $K_T = 1$ for the model examples $A = a\frac{\partial}{\partial x}$ and $A = \frac{\partial^2}{\partial x^2}$.

Assumption 6.10. *Let $\{T(t) \in \mathcal{L}(B, B) : t \geq 0\}$ be a semigroup with neutral element $T(0)$. Moreover, $T(t)$ is supposed to be uniformly bounded on $I = [0, T]$:*

$$\|T(t)\|_{B \leftarrow B} \leq K_T \qquad \text{for all } t \in I = [0, T]. \tag{6.8}$$

So far, only the homogeneous equation $u_t(t) = Au(t)$ has been studied.

Remark 6.11. *The solution of the inhomogeneous equation $\frac{\partial}{\partial t}u(t) = Au(t) + f(t)$ can be represented by*

$$u(t) = T(t)u_0 + \int_0^t T(t - s)f(s)\mathrm{d}s.$$

6.3 Discretisation of the Partial Differential Equation

6.3.1 Notations

We replace the real axis $\mathbb{R}$ of the x variable by an infinite grid of step size $\Delta x > 0$:

$$G_{\Delta x} = \{x = \nu \Delta x : \nu \in \mathbb{Z}\}. \tag{6.9}$$

Correspondingly, the interval $I = [0, T]$ is replaced by a finite grid of step size $\Delta t > 0$:

$$I_{\Delta t} = \{t = \mu \Delta t \leq T : \mu \in \mathbb{N}_0\}.$$

The Cartesian product of both grids yields the rectangular grid

$$\Sigma_{\Delta x}^{\Delta t} := I_{\Delta t} \times G_{\Delta x} = \{(t, x) \in \Sigma : x/\Delta x \in \mathbb{Z}, \ t/\Delta t \in \mathbb{N}_0\} \tag{6.10}$$

(cf. (6.3)). As we shall see, the step sizes Δx, Δt are, in general, not chosen independently, but are connected by a parameter λ (the power of Δx corresponds to the order of the differential operator A):

$$\lambda = \begin{cases} \Delta t/\Delta x & \text{in the hyperbolic case } A = a\frac{\partial}{\partial x}, \\ \Delta t/\Delta x^2 & \text{in the parabolic case } A = \frac{\partial^2}{\partial x^2}. \end{cases} \tag{6.11}$$

For a *grid function* $U : \Sigma_{\Delta x}^{\Delta t} \to \mathbb{C}$ we use the notation

$$U_\nu^\mu := U(\mu\Delta t, \nu\Delta x) \qquad \text{for } (\mu\Delta t, \nu\Delta x) \in \Sigma_{\Delta x}^{\Delta t}, \ \mu \in \mathbb{N}_0, \ \nu \in \mathbb{Z}.$$

All grid values at $t = \mu\Delta t$ are collected in

$$U^\mu := (U_\nu^\mu)_{\nu\in\mathbb{Z}}.$$

Let $\ell = \mathbb{C}^\mathbb{Z}$ be the linear space of two-sided infinite sequences with componentwise addition and multiplication by scalars. The Banach spaces $B = C(\mathbb{R})$ (or $L^\infty(\mathbb{R})$) and $B = L^2(\mathbb{R})$ correspond to the sequence spaces ℓ^∞ and ℓ^2:

- $\ell^\infty = \mathbb{C}^\mathbb{Z}$ with the norm $\|U\|_{\ell^\infty} = \sup\{|U_\nu| : \nu \in \mathbb{Z}\}$ is a Banach space,
- $\ell^2 = \mathbb{C}^\mathbb{Z}$ with the norm $\|U\|_{\ell^2} = \sqrt{\Delta x \sum_{\nu\in\mathbb{Z}} |U_\nu|^2}$ is a Hilbert space.

As a common symbol we use ℓ^p ($p \in \{2, \infty\}$).

6.3.2 Transfer Operators r, p

The continuous Banach space B and the discrete space ℓ^p of grid functions are connected via

$$r = r_{\Delta x} : B \to \ell^p. \tag{6.12a}$$

The letter r means 'restriction'. The index Δx will be omitted, when the underlying step size is known.

In the case of $B = C(\mathbb{R})$, an obvious choice of r is the evaluation at the grid points of $G_{\Delta x}$:

$$u \in C(\mathbb{R}) \Rightarrow ru \in \ell^\infty \text{ with } (ru)_j = u(j\Delta x) \quad \text{for } j \in \mathbb{Z}. \tag{6.12b}$$

In the case of $B = L^2(\mathbb{R})$ this is impossible, since L^2 functions have no well-defined point evaluations. Instead, one can use mean values:

$$u \in L^2(\mathbb{R}) \Rightarrow ru \in \ell^2 \text{ with } (ru)_j = \frac{1}{\Delta x} \int_{(j-1/2)\Delta x}^{(j+1/2)\Delta x} u(x)\mathrm{d}x \quad \text{for } j \in \mathbb{Z}.$$
$$\tag{6.12c}$$

We suppose that $r = r_{\Delta x} : B \to \ell^p$ is bounded:

$$\|r_{\Delta x}\|_{\ell^p \leftarrow B} \leq C_r \qquad \text{for all } \Delta x > 0. \tag{6.13}$$

Lemma 6.12. *The restrictions (6.12b,c) satisfy condition (6.13) with $C_r = 1$ with respect to the respective norms $\|\cdot\|_{\ell^\infty \leftarrow C(\mathbb{R})}$ and $\|\cdot\|_{\ell^2 \leftarrow L^2(\mathbb{R})}.$*

Proof. (i) $\|ru\|_{\ell^\infty} = \sup\{|u(j\Delta x)| : j \in \mathbb{Z}\} \le \sup\{|u(x)| : x \in \mathbb{R}\} = \|u\|_{C(\mathbb{R})}$ yields (6.12b).

(ii) $\|ru\|_{\ell^2}^2 = \|ru\|_2^2 = \Delta x \sum_j |(ru)_j|^2 = \frac{1}{\Delta x} \sum_j \left| \int_{(j-1/2)\Delta x}^{(j+1/2)\Delta x} u(x)\mathrm{d}x \right|^2$ and Schwarz' inequality

$$\left| \int_{(j-1/2)\Delta x}^{(j+1/2)\Delta x} u(x)\mathrm{d}x \right|^2 \le \left[\int_{(j-1/2)\Delta x}^{(j+1/2)\Delta x} 1 \mathrm{d}x \right] \cdot \left[\int_{(j-1/2)\Delta x}^{(j+1/2)\Delta x} |u(x)|^2 \mathrm{d}x \right]$$

$$= \Delta x \int_{(j-1/2)\Delta x}^{(j+1/2)\Delta x} |u(x)|^2 \mathrm{d}x$$

yield (6.12c): $\|ru\|_{\ell^2}^2 \le \sum_j \int_{(j-1/2)\Delta x}^{(j+1/2)\Delta x} |u(x)|^2 \mathrm{d}x = \int_{\mathbb{R}} |u(x)|^2 \mathrm{d}x = \|u\|_{L^2(\mathbb{R})}^2.$ □

The 'prolongation' p acts in the reverse direction: $p = p_{\Delta x} : \ell^p \to B$. We suppose the existence of a prolongation p with the following properties:

$$\|p_{\Delta x}\|_{B \leftarrow \ell^p} \le C_p \text{ for all } \Delta x > 0 \qquad \text{and} \qquad rp = I. \qquad (6.14)$$

The condition $rp = I$ indicates that p is a right-inverse of r.

Exercise 6.13. *Verify that in the cases (6.12b,c) the following choices of p satisfy condition (6.14) with $C_p = 1$:*

        p is piecewise linear interpolation: (6.15a)

        $v \in \ell^\infty \mapsto pv \in C(\mathbb{R})$ with $(pv)(x) = \vartheta v_j + (1 - \vartheta) v_{j+1},$

        where $x = (j + \vartheta)\Delta x$, $j \in \mathbb{Z}$, $\vartheta \in [0, 1),$

or

        p is piecewise constant interpolation: (6.15b)

        $v \in \ell^2 \mapsto pv \in L^2(\mathbb{R})$ with $(pv)(x) = v_j,$

        where $x \in \left[(j - \frac{1}{2})\Delta x, (j + \frac{1}{2})\Delta x\right), j \in \mathbb{Z}.$

6.3.3 Difference Schemes

Initially, U^0 is prescribed via $U_\nu^0 = ru_0$ (r from (6.12a), u_0 from (6.1b)). The *explicit* difference scheme

$$U_\nu^{\mu+1} = \sum_{j \in \mathbb{Z}} a_j U_{\nu+j}^\mu \qquad (6.16)$$

allows us to compute the next time step $U^{\mu+1}$ from U^μ. In practice, $\sum_{j\in\mathbb{Z}}$ is a finite sum; i.e., almost all a_j vanish.

Example 6.14. *(a) In the* hyperbolic *case, one may replace $\frac{\partial}{\partial t}u$ by the difference quotient $\frac{u(t+\Delta t,x)-u(t,x)}{\Delta t}$ and $A = a\frac{\partial u}{\partial x}$ by $a\frac{u(t,x+\Delta x)-u(t,x)}{\Delta x}$. Solving*

$$\frac{u(t+\Delta t, x) - u(t, x)}{\Delta t} = a\frac{u(t, x+\Delta x) - u(t, x)}{\Delta x}$$

with respect to $u(t+\Delta t, x)$ and using the relation (6.11), one obtains

$$u(t+\Delta t, x) = u(t, x) + \frac{a\Delta t}{\Delta x}\left[u(t, x+\Delta x) - u(t, x)\right];$$

$$i.e., \quad U_\nu^{\mu+1} = (1 - a\lambda)\, U_\nu^\mu + a\lambda U_{\nu+1}^\mu. \tag{6.17a}$$

(b) Replacing the right-sided difference $\frac{u(t,x+\Delta x)-u(t,x)}{\Delta x}$ by the symmetric difference $\frac{u(t,x+\Delta x)-u(t,x-\Delta x)}{2\Delta x}$, one derives the difference scheme

$$U_\nu^{\mu+1} = -\frac{a\lambda}{2}U_{\nu-1}^\mu + U_\nu^\mu + \frac{a\lambda}{2}U_{\nu+1}^\mu. \tag{6.17b}$$

(c) A replacement of $\frac{u(t+\Delta t,x)-u(t,x)}{\Delta t}$ by $\frac{u(t+\Delta t,x)-[u(t,x+\Delta x)+u(t,x-\Delta x)]/2}{\Delta t}$ yields

$$U_\nu^{\mu+1} = \frac{1 - a\lambda}{2}U_{\nu-1}^\mu + \frac{1 + a\lambda}{2}U_{\nu+1}^\mu. \tag{6.17c}$$

(d) In the parabolic *case of $A = \frac{\partial^2}{\partial x^2}$, the difference quotient $\frac{u(t+\Delta t,x)-u(t,x)}{\Delta t}$ for $\frac{\partial}{\partial t}u$ and the second difference quotient $\frac{u(t,x-\Delta x)-2u(t,x)+u(t,x+\Delta x)}{\Delta x^2}$ for $\frac{\partial^2 u}{\partial x^2}$ are obvious choices and lead together with $\lambda = \Delta t/\Delta x^2$ from (6.11) to*

$$\frac{u(t+\Delta t, x) - u(t, x)}{\Delta t} = \frac{2u(t, x) - u(t, x-\Delta x) - u(t, x+\Delta x)}{\Delta x^2};$$

i.e.,

$$U_\nu^{\mu+1} = \lambda U_{\nu-1}^\mu + (1 - 2\lambda)\, U_\nu^\mu + \lambda U_{\nu+1}^\mu. \tag{6.18}$$

The difference scheme (6.16) describes a linear mapping $U^\mu \mapsto U^{\mu+1}$ and defines the linear operator

$$C : \ell^p \to \ell^p, \quad (CU)_\nu := \sum_{j\in\mathbb{Z}} a_j U_{\nu+j} \text{ for } U \in \ell^p. \tag{6.19}$$

Hence, a shorter form of (6.16) is

$$U^{\mu+1} = CU^\mu. \tag{6.20}$$

Notation 6.15. *The difference operator C (and therefore also the coefficients a_j) may depend on the parameter λ and the step size Δt:*

$$C = C(\lambda, \Delta t).$$

The dependence on Δx follows implicitly via (6.11).

An example of a Δt-depending difference operator is given next.

Example 6.16. *Let $C(\lambda)$ be a suitable difference operator for $A = a\frac{\partial}{\partial x}$. Then the differential operator $A = a\frac{\partial}{\partial x} + b$ can be discretised by*

$$C'(\lambda, \Delta t) := C(\lambda) + \Delta t \cdot b;$$

i.e., a_0 from (6.16) is replaced by

$$a_0' := a_0 + \Delta t \cdot b.$$

Exercise 6.17. *Prove: (a) If $\sum_{j \in \mathbb{Z}} |a_j| < \infty$, then $C(\lambda, \Delta t) \in \mathcal{L}(\ell^\infty, \ell^\infty)$.*
(b) If $\sup_{j \in \mathbb{Z}} |a_j| < \infty$, then $C(\lambda, \Delta t) \in \mathcal{L}(\ell^2, \ell^2)$.

A μ-fold application of $C(\lambda, \Delta t)$ yields

$$U^\mu = C^\mu U^0 \quad \text{and} \quad U(\mu \Delta t, \nu \Delta x) = \left(C^\mu U^0\right)_\nu.$$

Remark 6.18. *So far, the coefficients a_j are assumed to be scalars. If the scalar equation $\frac{\partial}{\partial t} u = Au$ for $u : I \times \mathbb{R} \to \mathbb{R}$ is replaced by a vector-valued equation for $u : I \times \mathbb{R} \to \mathbb{R}^N$, the coefficients a_j in (6.19) become $N \times N$ matrices. The vector-valued case will be discussed in §6.5.5.*

The *shift operator E_j* is defined by

$$(E_j U)_\nu := U_{j+\nu}.$$

Since a shift does not change the ℓ^p-norm of U,

$$\|E_j\|_{\ell^p \leftarrow \ell^p} = 1 \tag{6.21}$$

holds. The difference operator C from (6.19) can be reformulated as

$$C := \sum_{j \in \mathbb{Z}} a_j E_j. \tag{6.22}$$

We conclude with a numerical computation of the example (6.17b), where we choose $a = 2$ and $\lambda = 1$; i.e., $\Delta t = \Delta x$. The initial value U^0 is the constant value 1 with the exception that at $x = 0$ we perturb 1 by 1.001. Without this perturbation, the solution of $u_t = 2u_x$ is $u(t, x) = 1$ and also the discrete solution is $U_\nu^\mu = 1$. We aim at the solution $u(1, x)$ at $t = 1$. Three computations with the step sizes

$\Delta t = 1, \frac{1}{10}, \frac{1}{100}$ are shown. The required number of time steps is 1, 10, or 100, respectively. The last three rows of the table list the values of $U_\nu^{1/\Delta t} = U(1, \nu \Delta x)$ for $-2 \le \nu \le 2$.

	$u(t, -2\Delta x)$	$u(t, -\Delta x)$	$u(t, 0)$	$u(t, \Delta x)$	$u(t, 2\Delta x)$	$\Delta t = \Delta x$
$t = 0$	1.000	1.000	1.001	1.000	1.000	
$t = 1$	1.000	1.001	1.001	0.999	1.000	1
$t = 1$	1.045	0.230	0.869	1.770	1.045	1/10
$t = 1$	$6.298_{10}30$	$-3.754\,6_{10}30$	6462.5	$3.754\,6_{10}30$	$6.298_{10}30$	1/100

Instead of convergence to $u(t, x)$, we observe that completely wrong approximations are produced. The reason is an obvious instability of scheme (6.17b) (cf. Example 6.46b). Note that we may replace the perturbed value 1.001 also by $1 + 10^{-16}$. This would change, e.g., $6.298_{10}30$ to $6.298_{10}17$, which is equally wrong.

6.4 Consistency, Convergence, and Stability

6.4.1 Definitions

Again, the restriction r from §6.3.2 is used. Consistency will depend on the underlying norm ℓ^p, where $p = 2, \infty$ are the discussed examples corresponding to the Banach spaces $B = L^2(\mathbb{R})$ and $C(\mathbb{R})$. We say that 'ℓ^p is suited to B', if the requirements (6.13) and (6.14) are satisfied.

Definition 6.19 (consistency). *Let $B_0 \subset D_A$ be a dense subset of B. Let ℓ^p be suited to B. The solution of (6.1a,b) is denoted by $u(t) = T(t)u_0$, $u_0 \in B_0$. The local discretisation error τ is defined by*

$$\tau(t) = \frac{1}{\Delta t} \left[r u(t + \Delta t) - C(\lambda, \Delta t) r u(t) \right].$$

The difference scheme $C(\lambda, \Delta t)$ is called consistent *(with respect to ℓ^p) if for all $u_0 \in B_0$,*

$$\sup \left\{ \|\tau(t)\|_{\ell^p} : 0 \le t \le T - \Delta t \right\} \to 0 \qquad \text{as } \Delta t \to 0.$$

The latter condition is equivalent to

$$\sup_{0 \le t \le T - \Delta t} \|[r T(\Delta t) - C(\lambda, \Delta t) r] \, T(t) u_0\|_{\ell^p} := o(\Delta t) \quad \text{for all } u_0 \in B_0.$$

Note that the following definition of convergence refers to the whole Banach space B, not to a dense subspace B_0.

Definition 6.20 (convergence). *For all $u_0 \in B$, let $u(t) = T(t)u_0$ denote the generalised solution. The difference scheme $C(\lambda, \Delta t)$ is called* convergent *(with respect to ℓ^p), if*

$$\|ru(t) - C(\lambda, \Delta t)^\mu r u_0\|_{\ell^p} \to 0 \qquad \text{for } \Delta t \to 0 \text{ and } \mu \Delta t \to t \in I = [0, T].$$

Definition 6.21 (stability). *The difference scheme $C(\lambda, \Delta t)$ is called* stable *(with respect to ℓ^p) if*

$$\sup\{\|C(\lambda, \Delta t)^\mu\|_{\ell^p \leftarrow \ell^p} : \Delta t \geq 0, \ \mu \in \mathbb{N}_0, \ 0 \leq \mu \Delta t \leq T\} < \infty. \qquad (6.23)$$

If (6.23) holds only for certain values of λ, the scheme is called conditionally stable; *otherwise, the scheme is called* unconditionally stable.

In the case of (6.23), the *stability constant* is defined by

$$K = K(\lambda) := \sup\{\|C(\lambda, \Delta t)^\mu\|_{\ell^p \leftarrow \ell^p} : \Delta t \geq 0, \ \mu \in \mathbb{N}_0, \ 0 \leq \mu \Delta t \leq T\}.$$

Instead of 'stable with respect to ℓ^p' we say for short 'ℓ^p stable'. Similarly, the terms 'ℓ^p stability', 'ℓ^p consistent', 'ℓ^p consistency' etc. are used.

6.4.2 Convergence, Stability and Equivalence Theorems

First, we show that consistency and stability—together with some mild technical assumptions—imply convergence.

Theorem 6.22 (convergence theorem). *Suppose (a) to (d):*
  (a) *r bounded with respect to ℓ^p (cf. (6.13)),*
  (b) *$T(t)$ satisfies Assumption 6.10,*
  (c) *ℓ^p stability of the difference scheme $C(\lambda, \Delta t)$,*
  (d) *ℓ^p consistency.*
Then the difference scheme is convergent with respect to ℓ^p.

Proof. (i) Given an initial value $u_0 \in B_0 \subset D_A$ (B_0 dense subset of B), define $u(t) = T(t)u_0$. We split the discretisation error as follows:

$$ru(t) - C(\lambda, \Delta t)^\mu r u_0 = r\left[u(t) - u(\mu \Delta t)\right] + \left[rT(\mu \Delta t) - C(\lambda, \Delta t)^\mu r\right] u_0$$
$$= r\left[u(t) - u(\mu \Delta t)\right] + \left[rT(\Delta t)^\mu - C(\lambda, \Delta t)^\mu r\right] u_0.$$

Use the telescopic sum

$$rA^\mu - B^\mu r = \sum_{\nu=0}^{\mu-1} B^\nu \left[rA - Br\right] A^{\mu-\nu-1}$$

with $A := T(\Delta t)$ and $B := C(\lambda, \Delta t)$. This shows

$$\|ru(t) - C(\lambda, \Delta t)^\mu r u_0\|_{\ell^p} \leq \|r\|_{\ell^p \leftarrow B} \|u(t) - u(\mu \Delta t)\|_B +$$

$$+ \sum_{\nu=0}^{\mu-1} \|C(\lambda, \Delta t)^\nu\|_{\ell^p \leftarrow \ell^p} \left\|\{rT(\Delta t) - C(\lambda, \Delta t)r\}u\big((\mu - \nu - 1)\Delta t\big)\right\|_{\ell^p}$$

$$\leq K_r \, \|u(t) - u(\mu\Delta t)\|_B + \sum_{\nu=0}^{\mu-1} K(\lambda)\Delta t \, \|\tau\big((\mu - \nu - 1)\,\Delta t\big)\|_{\ell^p}$$

with the stability constant $K(\lambda)$. Since $u_0 \in B_0 \subset D_A$, the solution $u(t) = T(t)u_0$ is strong, therefore differentiable and in particular continuous (cf. (6.7e)): $\|u(t) - u(\mu\Delta t)\|_B \to 0$ as $\mu\Delta t \to t$. Therefore, the first term is a zero sequence.

Because of the consistency assumption, the local discretisation error τ tends to zero uniformly. From $\|\tau((\mu - \nu - 1)\,\Delta t)\|_{\ell^p} \leq \varepsilon$ we infer that

$$\sum_{\nu=0}^{\mu-1} K(\lambda)\Delta t \, \|\tau((\mu - \nu - 1)\,\Delta t)\|_{\ell^p} \leq \mu K(\lambda)\Delta t\varepsilon \underset{\mu\Delta t\leq T}{\leq} K(\lambda)T\varepsilon,$$

so that the whole sum tends to zero. This proves convergence in the case of an initial value $u_0 \in B_0$.

(ii) For a general initial value $u_0 \in B$ and an arbitrary $\varepsilon > 0$, one finds a $u_0^* \in B_0$ with $\|u_0 - u_0^*\|_B \leq \varepsilon / [3K_r \max\{\|T(t)\|_{B\leftarrow B}, K(\lambda)\}]$. The associated solution $u^*(t) = T(t)u_0^*$ satisfies

$$\|r\,[u(t) - u^*(t)]\|_{\ell^p} \leq K_r \, \|T(t)\,[u_0 - u_0^*]\|_B \leq K_r \, \|T(t)\|_{B\leftarrow B} \, \|u_0 - u_0^*\|_B \leq \frac{\varepsilon}{3}$$

and

$$\|C(\lambda, \Delta t)^\mu r u_0 - C(\lambda, \Delta t)^\mu r u_0^*\|_{\ell^p} = \|C(\lambda, \Delta t)^\mu r \,[u_0 - u_0^*]\|_{\ell^p}$$
$$\leq K(\lambda)K_r \, \|u_0 - u_0^*\|_B \leq \varepsilon/3.$$

Together with $\|r u^*(t) - C(\lambda, \Delta t)^\mu r u_0^*\|_{\ell^p} \leq \varepsilon/3$ from (i) for sufficiently small Δt and $t - \mu\Delta t$, it follows that $\|r u(t) - C(\lambda, \Delta t)^\mu r u_0\|_{\ell^p} \leq \varepsilon$, so that also for general initial values $u_0 \in B$ convergence is shown. $\square$

Next, we show that stability is also necessary for convergence.

Theorem 6.23 (stability theorem). *Choose B and ℓ^p suitably, so that (6.13), (6.14), and (6.8) hold. Then ℓ^p convergence implies ℓ^p stability.*

Proof. For an indirect proof assume that the difference scheme is unstable. Then there are sequences $\Delta t_\nu > 0, \mu_\nu \in \mathbb{N}_0$ with $0 \leq \mu_\nu \Delta t_\nu \leq T$, so that

$$\|C(\lambda, \Delta t_\nu)^{\mu_\nu}\|_{\ell^p \leftarrow \ell^p} \to \infty \qquad \text{for } \nu \to \infty.$$

Since the interval $I = [0, T]$ is compact, there is a subsequence with $\mu_\nu \Delta t_\nu \to t \in I$. The latter convergence yields

$$\|r u(t) - C(\lambda, \Delta t_\nu)^{\mu_\nu} r u_0\|_{\ell^p} \to 0 \qquad \text{for all } u_0 \in B,$$

and therefore,

$$\|C(\lambda, \Delta t_\nu)^{\mu_\nu} r u_0\|_{\ell^p} \leq 1 + \underset{u(t)=T(t)u_0}{\|r u(t)\|_{\ell^p} \leq} 1 + \|r\|_{\ell^p \leftarrow B} \, \|T(t)\|_{B\leftarrow B} \, \|u_0\|_B$$
$$\underset{(6.13),\,(6.8)}{\leq} 1 + K_r K_T \, \|u_0\|_B =: K_1(u_0) \qquad \text{for } \nu \geq \nu_0$$

with sufficiently large $\nu_0 = \nu_0(u_0)$. One concludes that $C_\nu := C(\lambda, \Delta t_\nu)^{\mu_\nu} r$ is a point-wise bounded sequence of operators. Corollary 3.39 yields that C_ν is uniformly bounded: there is a constant K with

$$\|C(\lambda, \Delta t_\nu)^{\mu_\nu} r\|_{\ell^p \leftarrow B} \leq K \qquad \text{for all } \nu \in \mathbb{N}.$$

Since, according to (6.14), p is a bounded right-inverse of r, it follows that

$$\begin{aligned}
\|C(\lambda, \Delta t_\nu)^{\mu_\nu}\|_{\ell^p \leftarrow \ell^p} &= \|C(\lambda, \Delta t_\nu)^{\mu_\nu} r p\|_{\ell^p \leftarrow \ell^p} \\
&\leq \|C(\lambda, \Delta t_\nu)^{\mu_\nu} r\|_{\ell^p \leftarrow B} \|p\|_{B \leftarrow \ell^p} \underset{(6.14)}{\leq} K K_p
\end{aligned}$$

in contradiction to the assumption $\|C(\lambda, \Delta t_\nu)^{\mu_\nu}\|_{\ell^p \leftarrow \ell^p} \to \infty$.   □

The conclusion from both theorems is the equivalence theorem.

Theorem 6.24 (equivalence theorem). *Suppose (6.8), (6.13), (6.14), and ℓ^p consistency. Then ℓ^p convergence and ℓ^p stability are equivalent.*

6.4.3 Other Norms

So far, we restrict the analysis to the ℓ^p and L^p norms for $p = 2$ and $p = \infty$. For $1 \leq p < \infty$, the L^p norm is defined in Exercise 6.4, while ℓ^p is defined analogously. The reason for the restriction to $p \in \{2, \infty\}$ is twofold. If different properties hold for ℓ^2 and ℓ^∞, a more involved analysis is necessary to describe the properties for ℓ^p with $2 < p < \infty$. It might be that the separation is between $p = 2$ and $p > 2$, because in the latter case the Hilbert structure is lost. However, it might also happen that properties change between the cases $p < \infty$ and $p = \infty$, since in the first case ℓ^p is reflexive, but not in the latter case.

If stability estimates hold for both cases $p = 2$ and $p = \infty$, we are in a very pleasant situation, since these bounds imply corresponding estimates for the ℓ^p and L^p setting for $2 < p < \infty$. This result is based on the interpolation estimate by Riesz–Thorin. It is proved by Marcel[7] Riesz [23] in[8] 1926/27, while Thorin [25] (1939) simplified the proof. In the following lemma, let $\|\cdot\|_{p \leftarrow p}$ be the operator norm of $\mathcal{L}(\ell^p, \ell^p)$ or $\mathcal{L}(L^p, L^p)$.

Lemma 6.25 (Riesz–Thorin theorem). *Assume $1 \leq p_1 \leq q \leq p_2 \leq \infty$. Then*

$$\|\cdot\|_{q \leftarrow q} \leq \|\cdot\|_{p_1 \leftarrow p_1}^{\alpha} \|\cdot\|_{p_2 \leftarrow p_2}^{\beta} \qquad \text{with } \alpha = \frac{q - p_1}{p_2 - p_1}, \ \beta = \frac{p_2 - q}{p_2 - p_1}.$$

---

[7] Marcel Riesz is the younger brother of Frigyes Riesz, who is the author of, e.g., [22].

[8] The article belongs to Volume 49, which is associated to the year 1926. However, it is the last article of that volume and carries the footnote 'Imprimé le 11 janvier 1927'. Therefore, one finds also 1927 as the publication date.

Note that $\alpha + \beta = 1$. This implies that p_1 stability (i.e., $\|\ldots\|_{p_1 \leftarrow p_1} \leq M_1$) and p_2 stability $\|\ldots\|_{p_2 \leftarrow p_2} \leq M_2$ imply q stability in the form

$$\|\ldots\|_{q \leftarrow q} \leq M := M_1^\alpha M_2^\beta.$$

Hence, if a criterion yields both ℓ^2 and ℓ^∞ stability, then ℓ^p stability holds for all $2 \leq p \leq \infty$.

6.5 Sufficient and Necessary Conditions for Stability

6.5.1 First Criteria

The following results belong to the classical stability theory of Lax–Richtmyer [16].

Criterion 6.26. *If*

$$\|C(\lambda, \Delta t)\|_{\ell^p \leftarrow \ell^p} \leq 1 + K_\lambda \Delta t \qquad \textit{for all } \Delta t > 0,$$

then the difference scheme is ℓ^p stable with stability constant $K(\lambda) := e^{TK_\lambda}$.

Proof. Use

$$\|C(\lambda, \Delta t)^\mu\|_{\ell^p \leftarrow \ell^p} \leq \|C(\lambda, \Delta t)\|_{\ell^p \leftarrow \ell^p}^\mu \leq (1 + K_\lambda \Delta t)^\mu$$

and apply Exercise 3.24a: $(1 + K_\lambda \Delta t)^\mu \leq \left(e^{K_\lambda \Delta t}\right)^\mu = e^{K_\lambda \mu \Delta t} \underset{\mu \Delta t \leq T}{\leq} e^{K_\lambda T}.$ $\square$

The coefficients a_j of $C(\lambda, \Delta t)$ can be used to estimate $\|C(\lambda, \Delta t)\|_{\ell^p \leftarrow \ell^p}$.

Remark 6.27. *The difference scheme (6.19) satisfies $\|C(\lambda, \Delta t)\|_{\ell^p \leftarrow \ell^p} \leq \sum |a_j|$.*

Proof. $\|C(\lambda, \Delta t)\|_{\ell^p \leftarrow \ell^p} = \left\|\sum_j a_j E_j\right\|_{\ell^p \leftarrow \ell^p} \leq \sum_j |a_j| \, \|E_j\|_{\ell^p \leftarrow \ell^p} = \sum_j |a_j|$ follows from (6.22) and (6.21). $\square$

Combining the previous statements, we arrive at the next result.

Corollary 6.28. *If $\sum_j |a_j| \leq 1 + K_\lambda \Delta t$ for all $\Delta t > 0$, the difference scheme is stable (with respect to the ℓ^2 and ℓ^∞ norms) with stability constant $K(\lambda) := e^{TK_\lambda}$.*

Definition 6.29 (positive difference scheme). *A difference scheme (6.19) is called positive if $a_j \geq 0$ holds for all coefficients.*

Positive difference schemes map non-negative initial values $U^0 \in \ell^p$ into non-negative solutions U^μ, which often is an appreciated property. From Corollary 6.28 we infer the next one.

Criterion 6.30. *Positive difference schemes (6.19) with $\sum_j a_j = 1 + \mathcal{O}(\Delta t)$ are stable (with respect to the ℓ^2 and ℓ^∞ norms).*

We can also formulate a criterion for instability.

Criterion 6.31. *Suppose that $\sum a_j \geq 1 + \Delta t\, c(\Delta t)$ with $\lim_{\Delta t \to 0} c(\Delta t) = \infty$. Then the difference scheme (6.19) is unstable (with respect to ℓ^2 and ℓ^∞ norms).*

Proof. (i) In the case of ℓ^∞, choose the constant grid function $U^0 \in \ell^\infty$, i.e., $U_j^0 = 1$ for all $j \in \mathbb{Z}$. Then $U^1 = C(\lambda, \Delta t)U^0$ equals ζU^0 with $\zeta := \sum a_j$ and, correspondingly, $U^\mu = C(\lambda, \Delta t)^\mu U^0 = \zeta^\mu U^0$. Hence, $\|C(\lambda, \Delta t)^\mu\|_{\ell^p \leftarrow \ell^p} \geq \zeta^\mu \geq [1 + \Delta t c(\Delta t)]^\mu$. Exercise 6.32a proves the assertion.

(ii) In the case of ℓ^2, the previous proof cannot be repeated, since $U^0 \notin \ell^2$. Instead, the proof will be given after Theorem 6.44 on page 118.   □

Exercise 6.32. *Suppose that $\lim_{\Delta t \to 0} c(\Delta t) = \infty$. (a) Prove*

$$\sup\{[1 + \Delta t\, c(\Delta t)]^\mu : \mu \in \mathbb{N}_0,\ \Delta t > 0,\ \mu \Delta t \leq T\} = \infty.$$

(b) Show that for any constant $K > 0$ and all positive integers μ with $\mu \leq T/\Delta t$, the μ-th root $\sqrt[\mu]{K}$ satisfies the inequality $\sqrt[\mu]{K} \leq 1 + C_\rho \Delta t$. Hint: set $c(\Delta t) := \{\sqrt[\mu]{K} - 1\}/\Delta t$ and prove that $C_\rho := \sup_{\Delta t > 0} c(\Delta t) < \infty$.

Next, we apply the criteria to the examples in §6.3.3.

Example 6.33. *(a) In (6.17a), the non-zero coefficients are $a_0 = 1 - a\lambda$ and $a_1 = a\lambda$. Assume $a \geq 0$. For λ with $0 \leq a\lambda \leq 1$, (6.17a) is a positive scheme with $\sum a_j = 1$; hence, it is ℓ^2 and ℓ^∞ stable according to Corollary 6.28.*

(b) In (6.17b), the non-zero coefficients are $a_{-1} = -\frac{a\lambda}{2}$, $a_0 = 1$, and $a_1 = \frac{a\lambda}{2}$. Excluding the trivial case $a = 0$ (cf. Footnote 4 on page 95), we cannot apply Corollary 6.28, since $\sum |a_j| = 1 + a\lambda > 1$. (6.17b) is not a positive scheme.

(c) In (6.17c), the non-zero coefficients are $a_{-1} = \frac{1 - a\lambda}{2}$ and $a_1 = \frac{1 + a\lambda}{2}$. Under the condition $|a\lambda| \leq 1$ the scheme is positive, and Criterion 6.30 ensures ℓ^2 and ℓ^∞ stability: $\sum a_j = 1$.

(d) In (6.18), the non-zero coefficients are $a_{-1} = \lambda$, $a_0 = 1 - 2\lambda$, $a_1 = \lambda$. For $\lambda \in (0, 1/2]$ the scheme is positive, and Criterion 6.30 ensures ℓ^2 and ℓ^∞ stability.

An interesting question is whether a stable scheme remains stable after a perturbation (cf. §3.5.3).

Lemma 6.34 (perturbation lemma). *Let $C(\lambda, \Delta t)$ be ℓ^p stable with stability constant $K(\lambda)$. Suppose that a perturbation $D(\lambda, \Delta t)$ is bounded by*

$$\|D(\lambda, \Delta t)\|_{\ell^p \leftarrow \ell^p} \leq C_D \Delta t.$$

Then

$$C'(\lambda, \Delta t) := C(\lambda, \Delta t) + D(\lambda, \Delta t)$$

is again ℓ^p stable with stability constant [9]

$$K'(\lambda) \leq K(\lambda)e^{K(\lambda)C_D T},$$

where $I = [0,T]$ is the given time interval. This result holds also in the case that $C(\lambda, \Delta t)$ and $D(\lambda, \Delta t)$ are non-commutative [10] *operators.*

Proof. We have to estimate $C'(\lambda, \Delta t)^\mu$. The simple binomial formula is only valid for commutative terms. In the general case, we have

$$[C + D]^\mu = \sum_{m=0}^{\mu} \sum_{\alpha_1 + \ldots + \alpha_{\mu+1} = \mu - m} C^{\alpha_1} D C^{\alpha_2} D \cdots C^{\alpha_m} D C^{\alpha_{m+1}},$$

where the second sum runs over all $\alpha_j \in \mathbb{N}_0$ with $\alpha_1 + \ldots + \alpha_{\mu+1} = \mu - m$. Each term $C^{\alpha_1} D C^{\alpha_2} D \cdots C^{\alpha_m} D C^{\alpha_{m+1}}$ contains m factors D and $\mu - m$ factors C. For a fixed $m \in [0, \mu]$ the number of these terms is $\binom{\mu}{m}$. Together with the estimate

$$\|C^{\alpha_1} D C^{\alpha_2} D \cdots C^{\alpha_\mu} D C^{\alpha_{\mu+1}}\|$$
$$\leq \|C^{\alpha_1}\| \|D\| \|C^{\alpha_2}\| \|D\| \cdots \|C^{\alpha_m}\| \|D\| \|C^{\alpha_{m+1}}\| \leq K(\lambda)^{m+1} (C_D \Delta t)^m$$

we arrive at the inequality

$$\|C'(\lambda, \Delta t)^\mu\|_{\ell^p \leftarrow \ell^p} =$$
$$= \|[C(\lambda, \Delta t) + D(\lambda, \Delta t)]^\mu\|_{\ell^p \leftarrow \ell^p} \leq \sum_{m=0}^{\mu} \binom{\mu}{m} K(\lambda)^{m+1} (C_D \Delta t)^m$$
$$= K(\lambda) \sum_{m=0}^{\mu} \binom{\mu}{m} (K(\lambda)C_D \Delta t)^m = K(\lambda) [1 + K(\lambda)C_D \Delta t]^\mu.$$

with $[1 + K(\lambda)C_D \Delta t]^\mu \leq e^{K(\lambda)C_D \mu \Delta t} \underset{\mu \Delta t \leq T}{\leq} e^{K(\lambda)C_D T}$ (cf. Exercise 3.24). □

Remark 6.35. *(a) A simple application of Lemma 6.34 is the following one. Let the differential operator A in $\frac{\partial}{\partial t}u = Au$ (cf. (6.1a)) be $A = A_1 + A_0$, where A_1 contains derivatives of at least first order, while $A_0 u = a_0 u$ is the term of order zero. The discretisation yields correspondingly $C(\lambda, \Delta t) = C_1(\lambda, \Delta t) + C_0(\lambda, \Delta t)$. A consistent discretisation of C_0 satisfies the estimate $\|C_0(\lambda, \Delta t)\|_{\ell^p \leftarrow \ell^p} = \mathcal{O}(\Delta t)$. By Lemma 6.34, the stability of $C_1(\lambda, \Delta t)$ implies the stability of $C(\lambda, \Delta t)$. Therefore, it suffices to investigate differential operators A without terms of order zero.*

(b) Let $A = A_1$ be the principal part from above. The property $A1 = 0$ ($1 \in \ell^\infty$: constant function with value one) shows that $u = 1$ is a solution. This implies the special consistency condition

$$\sum_{j \in \mathbb{Z}} a_j = 1 \qquad (= I \text{ in the matrix-valued case}) \tag{6.24}$$

for the coefficients of $C(\lambda, \Delta t)$ (cf. (6.19)).

---

[9] If commutativity $CD = DC$ holds, the stability constant improves: $K'(\lambda) \leq K(\lambda)e^{C_D T}$.

[10] Non-commutativity occurs for matrix-valued coefficients a_j (cf. §6.5.5).

Next we need the *spectral radius*[11]

$$\rho(A) := \sup\{|\lambda| : \lambda \text{ singular value of } A\}. \tag{6.25}$$

Criterion 6.36. *A necessary condition for stability (with respect to ℓ^2 and ℓ^∞) is*

$$\rho(C(\lambda, \Delta t)) \le 1 + \mathcal{O}(\Delta t).$$

Proof. $\rho(A^\mu) = \rho(A)^\mu$ holds for $\mu \in \mathbb{N}$. On the other hand, $\rho(A) \le \|A\|$ is valid for any associated norm. Therefore, stability yields

$$\rho(C(\lambda, \Delta t))^\mu = \rho(C(\lambda, \Delta t)^\mu) \le \|C(\lambda, \Delta t)^\mu\|_{\ell^p \leftarrow \ell^p} \le K(\lambda)$$

for all $\mu, \Delta t$ with $\mu \Delta t \le T$. Exercise 6.32b proves $\rho(C(\lambda, \Delta t)) \le 1 + C_\rho \Delta t$. $\square$

Remark 6.37. *Suppose that the operator $C(\lambda, \Delta t) \in \mathcal{L}(\ell^2, \ell^2)$ is normal; i.e., $C(\lambda, \Delta t)$ commutes with the adjoint operator $C(\lambda, \Delta t)^*$. Then $\rho(C(\lambda, \Delta t)) = \|C(\lambda, \Delta t)\|_{\ell^2 \leftarrow \ell^2}$ holds and ℓ^2 stability is equivalent to $\rho(C(\lambda, \Delta t)) \le 1 + \mathcal{O}(\Delta t)$.*

Proof. (i) Let C be a normal operator. First we prove $\|C^2\|_{\ell^2 \leftarrow \ell^2} = \|C^*C\|_{\ell^2 \leftarrow \ell^2}$. From

$$\langle CCu, CCu \rangle_{\ell^2} = \langle C^*CCu, Cu \rangle_{\ell^2} = \langle CC^*Cu, Cu \rangle_{\ell^2} = \langle C^*Cu, C^*Cu \rangle_{\ell^2}$$

it follows that

$$\|C^2\|^2_{\ell^2 \leftarrow \ell^2} = \sup_{\|u\|_{\ell^2}=1} \langle CCu, CCu \rangle_{\ell^2} = \sup_{\|u\|_{\ell^2}=1} \langle C^*Cu, C^*Cu \rangle_{\ell^2} = \|C^*C\|^2_{\ell^2 \leftarrow \ell^2}.$$

Since also

$$\|C^*C\|_{\ell^2 \leftarrow \ell^2} = \sup_{\|u\|_{\ell^2}=\|v\|_{\ell^2}=1} \langle u, C^*Cv \rangle_{\ell^2} = \sup_{\|u\|_{\ell^2}=\|v\|_{\ell^2}=1} \langle Cu, Cv \rangle_{\ell^2} \underset{u=v}{\ge}$$

$$\ge \sup_{\|u\|_{\ell^2}=1} \langle Cu, Cu \rangle_{\ell^2} = \|C\|^2_{\ell^2 \leftarrow \ell^2},$$

$\|C\|^2_{\ell^2 \leftarrow \ell^2} \ge \|C^2\|_{\ell^2 \leftarrow \ell^2}$ is shown. Because of $\|C^2\|_{\ell^2 \leftarrow \ell^2} \le \|C\|^2_{\ell^2 \leftarrow \ell^2}$ (submultiplicativity of the operator norm), the equality $\|C^2\|_{\ell^2 \leftarrow \ell^2} = \|C^*C\|_{\ell^2 \leftarrow \ell^2}$ is proved. Analogously, $\|C\|^n_{\ell^2 \leftarrow \ell^2} = \|C^n\|_{\ell^2 \leftarrow \ell^2}$ follows for all $n = 2^k$ ($k \in \mathbb{N}$).

(ii) The characterisation $\rho(C) = \lim_{n \to \infty} \sqrt[n]{\|C^n\|_{\ell^2 \leftarrow \ell^2}}$ together with (i) proves $\rho(C) = \|C\|_{\ell^2 \leftarrow \ell^2}$.

(iii) According to Criterion 6.36, $\rho(C(\lambda, \Delta t)) \le 1 + \mathcal{O}(\Delta t)$ is necessary, while $\|C(\lambda, \Delta t)\|_{\ell^2 \leftarrow \ell^2} \le 1 + \mathcal{O}(\Delta t)$ is sufficient (cf. Criterion 6.26). Since $\rho(C(\lambda, \Delta t)) = \|C(\lambda, \Delta t)\|_{\ell^2 \leftarrow \ell^2}$, both inequalities are identical. $\square$

---

[11] λ is regular value of A, if $\lambda I - A$ is bijective and the inverse $(\lambda I - A)^{-1} \in L(B, B)$ exists. Otherwise, λ is a singular value of A. In the case of finite-dimensional vector spaces (i.e., for matrices), the terms 'singular value' and 'eigenvalue' coincide.

Since $\rho(C(\lambda, \Delta t)) \leq 1 + \mathcal{O}(\Delta t)$ from Remark 6.37 is relatively easy to check, one may ask whether a similar criterion holds for more general operators. For this purpose we introduce the 'almost normal operators': $C(\lambda, \Delta t)$ is *almost normal* if

$$\|C(\lambda, \Delta t)C(\lambda, \Delta t)^* - C(\lambda, \Delta t)^* C(\lambda, \Delta t)\|_{\ell^2 \leftarrow \ell^2} \leq M (\Delta t)^2 \|C(\lambda, \Delta t)\|_{\ell^2 \leftarrow \ell^2}^2$$

for some constant $M < \infty$.

Criterion 6.38 (Lax [15, p. 41]). *If $C(\lambda, \Delta t)$ is almost normal, then ℓ^2 stability is equivalent to the estimate $\|C(\lambda, \Delta t)\|_{\ell^2 \leftarrow \ell^2} \leq 1 + \mathcal{O}(\Delta t)$.*

Proof. (i) Since $\|C(\lambda, \Delta t)\|_{\ell^2 \leftarrow \ell^2} \leq 1 + \mathcal{O}(\Delta t)$ is sufficient according to Criterion 6.26, it only remains to prove the necessity. Below, we abbreviate $C(\lambda, \Delta t)$ by C.

(ii) First we prove by induction that $(C^*)^\mu C^\mu$ can be reordered into $(C^*C)^\mu$ in at most $\mu^2/2$ steps (one step consists of $C^*C \mapsto CC^*$). In the case of $\mu = 1$, no permutation is required and $0 \leq (1^2)/2$ proves the start of induction. Suppose the induction hypothesis for μ. Then

$$(C^*C)^{\mu+1} \underset{\mu^2/2 \text{ steps}}{\hookrightarrow} (C^*)^\mu C^\mu C^* C \underset{\mu \text{ steps}}{\hookrightarrow} (C^*)^{\mu+1} C^{\mu+1}$$

and $\frac{\mu^2}{2} + \mu \leq \frac{(\mu+1)^2}{2}$ prove the hypothesis.

(iii) Each interchange perturbs the operator norm at most by $M (\Delta t)^2 \|C\|_{\ell^2 \leftarrow \ell^2}^{2\mu}$. In the following formula the factors C_j, $1 \leq j \leq 2\mu$, are either C or C^*:

$$\|C_1 \cdots C_\nu CC^* C_{\nu+3} \cdots C_{2\mu} - C_1 \cdots C_\nu C^* C C_{\nu+3} \cdots C_{2\mu}\|_{\ell^2 \leftarrow \ell^2}$$
$$\leq \|C_1 \cdots C_\nu\|_{\ell^2 \leftarrow \ell^2} \|CC^* - C^*C\|_{\ell^2 \leftarrow \ell^2} \|C_{\nu+3} \cdots C_{2\mu}\|_{\ell^2 \leftarrow \ell^2}$$
$$\leq \|C\|_{\ell^2 \leftarrow \ell^2}^\nu \left[M (\Delta t)^2 \|C\|_{\ell^2 \leftarrow \ell^2}^2 \right] \|C\|_{\ell^2 \leftarrow \ell^2}^{2\mu-\nu-2} = M (\Delta t)^2 \|C\|_{\ell^2 \leftarrow \ell^2}^{2\mu}$$

for all $0 \leq \nu \leq 2\mu - 2$ (note that $\|C^*\|_{\ell^2 \leftarrow \ell^2} = \|C\|_{\ell^2 \leftarrow \ell^2}$).

(iv) Forming the supremum over all $u \in \ell^2$ with $\|u\|_{\ell^2} = 1$ in

$$\langle C^\mu u, C^\mu u \rangle_{\ell^2} = \langle u, (C^*)^\mu C^\mu u \rangle_{\ell^2}$$
$$\underset{\text{(ii), (iii)}}{\geq} \langle u, (C^*C)^\mu u \rangle_{\ell^2} - \frac{M}{2} (\mu \Delta t)^2 \|C\|_{\ell^2 \leftarrow \ell^2}^{2\mu} \|u\|_{\ell^2}^2,$$

the left-hand side becomes $\|C^\mu\|_{\ell^2 \leftarrow \ell^2}^2$, while the right-hand side yields the lower bound $\left[1 - \frac{M}{2} (\mu \Delta t)^2\right] \|C\|_{\ell^2 \leftarrow \ell^2}^{2\mu}$ because of

$$\sup \langle u, (C^*C)^\mu u \rangle_{\ell^2} = \|(C^*C)^\mu\|_{\ell^2 \leftarrow \ell^2} \underset{C^*C \text{ normal}}{=} \|C^*C\|_{\ell^2 \leftarrow \ell^2}^\mu = \|C\|_{\ell^2 \leftarrow \ell^2}^{2\mu}.$$

Together with stability, this shows that

$$\left[1 - \frac{M}{2} (\mu \Delta t)^2\right] \|C\|_{\ell^2 \leftarrow \ell^2}^{2\mu} \leq \|C^\mu\|_{\ell^2 \leftarrow \ell^2}^2 \leq K(\lambda)^2. \tag{6.26}$$

Restricting μ and Δt by $\mu \Delta t \leq \min\{1/\sqrt{M}, T\}$, we obtain $1 - \frac{M}{2}(\mu \Delta t)^2 \geq \frac{1}{2}$. Inequality (6.26) implies $\|C\|_{\ell^2 \leftarrow \ell^2} \leq \sqrt[2\mu]{2K(\lambda)^2}$. Exercise 6.32b shows that $\|C\|_{\ell^2 \leftarrow \ell^2} = \|C(\lambda, \Delta t)\|_{\ell^2 \leftarrow \ell^2} \leq 1 + \mathcal{O}(\Delta t)$. $\square$

The previous statements are connected to estimates of the kind

$$\|C(\lambda, \Delta t)\|_{\ell^2 \leftarrow \ell^2} \leq 1 \qquad \text{or} \qquad \|C(\lambda, \Delta t)\|_{\ell^2 \leftarrow \ell^2} \leq 1 + \mathcal{O}(\Delta t).$$

These inequalities are the most convenient conditions proving stability. However, even if $\|C(\lambda, \Delta t)\|_{\ell^2 \leftarrow \ell^2} \geq c > 1$, it may happen that the powers stay bounded ($\|C(\lambda, \Delta t)^\mu\|_{\ell^2 \leftarrow \ell^2} \leq$ const). In that case one may try to find an equivalent norm which behaves easier. Since the square $\|U\|_{\ell^2}^2 = \sum_{i \in \mathbb{Z}} |U_i|^2$ is a quadratic form, one may introduce another quadratic form $Q(U)$ such that

$$\frac{1}{K_1} \|U\|_{\ell^2}^2 \leq Q(U) \leq K_1 \|U\|_{\ell^2}^2 \qquad \text{for all } U \in \ell^2. \qquad (6.27)$$

This describes the equivalence of the norms $\|\cdot\|_{\ell^2}$ and $\sqrt{Q(\cdot)}$. The next lemma is a slight generalisation of the 'energy method' stated in [21, pages 139–140].

Lemma 6.39. *Let (6.27) be valid. Suppose that the growth of $Q(\cdot)$ for one time step $U^{\mu+1} = C(\lambda, \Delta t)U^\mu$ is limited by*

$$0 \leq Q(U^{\mu+1}) - Q(U^\mu) \leq K_2 \Delta t \left(\|U^\mu\|_{\ell^2}^2 + \|U^{\mu+1}\|_{\ell^2}^2 \right) + K_3 \Delta t. \qquad (6.28a)$$

Then, for $\Delta t \leq 1/(2K_1 K_2)$, the norms $Q(U^\mu)$ and $\|U^\mu\|_{\ell^2}$ stay bounded. More precisely,

$$\|U^\mu\|_{\ell^2}^2 \leq \left\{ (2K_1^2 - 1) \|U^0\|_{\ell^2}^2 + \frac{K_3}{2K_2} \right\} \left(\frac{1 + K\Delta t}{1 - K\Delta t} \right)^\mu - \frac{K_3}{2K_2} \quad \text{for } \mu \in \mathbb{N}. \qquad (6.28b)$$

Proof. We introduce the following abbreviations:

$$q_\mu := Q(U^\mu), \; s_\mu := \|U^\mu\|_{\ell^2}^2, \; K := K_1 K_2, \; L := K_1 K_3, \; c := \frac{1 + K\Delta t}{1 - K\Delta t}.$$

Since $s_\mu \leq K_1 q_\mu = K_1 \left[q_0 + \sum_{\ell=0}^{\mu-1} (q_{\ell+1} - q_\ell) \right] \leq K_1 \left[K_1 s_0 + \sum_{\ell=0}^{\mu-1} (q_{\ell+1} - q_\ell) \right]$, we infer from (6.28a) that

$$s_\mu \leq \left(K_1^2 - K_1 K_2 \Delta t \right) s_0 + 2K_1 K_2 \Delta t \sum_{\ell=0}^{\mu-1} s_\ell + K_1 K_2 \Delta t \, s_\mu + K_1 K_3 \mu \Delta t.$$

Using the abbreviations K and L, we get

$$(1 - K\Delta t)\, s_\mu \le \left(K_1^2 - K\Delta t\right) s_0 + 2K\Delta t \sum_{\ell=0}^{\mu-1} s_\ell + L\mu\Delta t \quad \text{for all } \mu \in \mathbb{N}.$$

$$(6.29a)$$

Note that $1 - K\Delta t \ge 1/2$ because of $\Delta t \le 1/(2K_1K_2)$. The induction hypothesis

$$s_\mu \le S_\mu := Ac^\mu - B \quad \text{with } B := \frac{K_3}{2K_2},\ A := \left(2K_1^2 - 1\right) s_0 + B \qquad (6.29b)$$

is easily verified for $\mu = 0$, since $K_1^2 \ge 1$ follows from (6.27):

$$S_0 = Ac^0 - B = \left(2K_1^2 - 1\right) s_0 + B - B \ge s_0.$$

Assume that $s_\ell \le S_\ell$ holds for all $\ell \le \mu - 1$. Insertion of $s_\ell \le S_\ell$ and (6.29b) into (6.29a) yields

$$(1 - K\Delta t)\, s_\mu \le \left(K_1^2 - K\Delta t\right) s_0 + 2K\Delta t \sum_{\ell=0}^{\mu-1} S_\ell + L\mu\Delta t$$

$$\le \left(K_1^2 - K\Delta t\right) s_0 + 2AK\Delta t \sum_{\ell=0}^{\mu-1} c^\ell - 2BK\mu\Delta t + L\mu\Delta t.$$

Using $\sum_{\ell=0}^{\mu-1} c^\ell = \frac{c^\mu - 1}{c - 1} = \frac{1 - K\Delta t}{2K\Delta t} (c^\mu - 1)$, we continue:

$$(1 - K\Delta t)\, s_\mu \le \left(K_1^2 - K\Delta t\right) s_0 + (1 - K\Delta t)\, A\, (c^\mu - 1) + (L - 2KB)\, \mu\Delta t$$

$$\underset{L=2KB}{=} (1 - K\Delta t) \Big\{ \underbrace{Ac^\mu - B}_{=S_\mu} + B + \tfrac{K_1^2 - K\Delta t}{1 - K\Delta t} s_0 - A \Big\}.$$

The expression $\frac{K_1^2 - K\Delta t}{1 - K\Delta t}$ is monotone in $K\Delta t$. From $K\Delta t \le 1/2$ we infer that

$$B + \tfrac{K_1^2 - K\Delta t}{1 - K\Delta t} s_0 - A \le B + \left(2K_1^2 - 1\right) s_0 - A = 0,$$

which proves $s_\mu \le S_\mu$ for μ. $\square$

To derive further stability criteria in the case of ℓ^2, we need Fourier transformations, which are provided in the next part.

6.5.2 Fourier Analysis

A 2π-periodic function $f \in L^2_{2\pi}(\mathbb{R})$ is the 2π-periodic extension of $f \in L^2(0, 2\pi)$ to $\mathbb{R}$. The associated Fourier series is

$$\frac{1}{\sqrt{2\pi}} \sum_{\alpha \in \mathbb{Z}} \varphi_\alpha e^{i\alpha\xi} \quad \text{with } \varphi_\alpha := \frac{1}{\sqrt{2\pi}} \int_0^{2\pi} f(\xi) e^{-i\alpha\xi} d\xi \quad (\alpha \in \mathbb{Z}),$$

where $i = \sqrt{-1}$. In the case of sufficiently smooth 2π-periodic functions f, the Fourier series converges uniformly (with respect to $\|\cdot\|_\infty$) to f. For general L^2 functions, the convergence holds only in the sense of the L^2 norm. Therefore, the series in $f(\xi) = \frac{1}{\sqrt{2\pi}} \sum_{\alpha \in \mathbb{Z}} \varphi_\alpha e^{i\alpha\xi}$ has to be understood in the sense of $\|\cdot\|_{L^2(0,2\pi)}$. In the sequel, the following isometry (Parseval equality) is essential:

$$\|f\|_{L^2(0,2\pi)} = \|\varphi\|_{\ell^2}, \qquad \text{where } \varphi = (\varphi_\alpha)_{\alpha \in \mathbb{Z}}. \tag{6.30}$$

The transfer from the function $f \in L^2(0, 2\pi)$ to its Fourier coefficients $\varphi \in \ell^2$ is the *Fourier analysis*, which will be denoted by $\mathcal{F}$:

$$\mathcal{F}f = \varphi.$$

On the other hand, the mapping $\varphi \mapsto \frac{1}{\sqrt{2\pi}} \sum_{\alpha \in \mathbb{Z}} \varphi_\alpha e^{i\alpha\xi} = f$ is called the *Fourier synthesis* and coincides with the inverse $\mathcal{F}^{-1}$.

For $p = 2$, the solutions U^μ of the difference scheme are ℓ^2 sequences. We denote the associated 2π-periodic functions by $\hat{U}^\mu$:

$$\hat{U}^\mu := \mathcal{F}^{-1} U^\mu, \qquad \hat{U}^\mu(\xi) = \frac{1}{\sqrt{2\pi}} \sum_{\alpha \in \mathbb{Z}} U_\alpha^\mu e^{i\alpha\xi}.$$

The difference scheme $U^{\mu+1} = C(\lambda, \Delta t) U^\mu$ is equivalent to

$$\hat{U}^{\mu+1} = \mathcal{F}^{-1} U^{\mu+1} = \mathcal{F}^{-1} C(\lambda, \Delta t) U^\mu = \mathcal{F}^{-1} C(\lambda, \Delta t) \mathcal{F} \mathcal{F}^{-1} U^\mu = \hat{C}(\lambda, \Delta t) \hat{U}^\mu$$

with

$$\hat{C}(\lambda, \Delta t) := \mathcal{F}^{-1} C(\lambda, \Delta t) \mathcal{F}. \tag{6.31}$$

While $C(\lambda, \Delta t) \in \mathcal{L}(\ell^2, \ell^2)$, the transformed operator $\hat{C}(\lambda, \Delta t)$ belongs to the space $\mathcal{L}(L^2(0, 2\pi), L^2(0, 2\pi))$.

Exercise 6.40. *Using the isometry (6.30), prove*

$$\|\mathcal{F}\|_{\ell^2 \leftarrow L^2(0,2\pi)} = \|\mathcal{F}^{-1}\|_{L^2(0,2\pi) \leftarrow \ell^2} = 1 \tag{6.32}$$

(i.e., $\mathcal{F}$ and $\mathcal{F}^{-1}$ are unitary) and show that

$$\|A\|_{\ell^2 \leftarrow \ell^2} = \|\mathcal{F}^{-1} A\|_{L^2(0,2\pi) \leftarrow \ell^2} \qquad \text{for all } A \in \mathcal{L}(\ell^2, \ell^2),$$
$$\|B\|_{L^2(0,2\pi) \leftarrow \ell^2} = \|B\mathcal{F}\|_{L^2(0,2\pi) \leftarrow L^2(0,2\pi)} \quad \text{for all } B \in \mathcal{L}(\ell^2, L^2(0, 2\pi)).$$

The decisive quantity for stability is the operator norm $\|C(\lambda, \Delta t)^\mu\|_{\ell^2 \leftarrow \ell^2}$. The previous exercise shows that

$$\|C(\lambda, \Delta t)^\mu\|_{\ell^2 \leftarrow \ell^2} = \|\mathcal{F}^{-1} C(\lambda, \Delta t)^\mu\|_{L^2(0,2\pi) \leftarrow \ell^2}$$
$$= \|\mathcal{F}^{-1} C(\lambda, \Delta t)^\mu \mathcal{F}\|_{L^2(0,2\pi) \leftarrow L^2(0,2\pi)} = \|\left[\mathcal{F} C(\lambda, \Delta t) \mathcal{F}^{-1}\right]^\mu\|_{L^2(0,2\pi) \leftarrow L^2(0,2\pi)}$$
$$= \|\hat{C}(\lambda, \Delta t)^\mu\|_{L^2(0,2\pi) \leftarrow L^2(0,2\pi)}. \tag{6.33}$$

This leads us to the following remark.

Remark 6.41. *An equivalent definition of ℓ^2 stability with the stability constant $K(\lambda)$ is*

$$\|\hat{C}(\lambda, \Delta t)^\mu\|_{L^2(0,2\pi)\leftarrow L^2(0,2\pi)} \leq K(\lambda) \quad \text{for all } \mu \in \mathbb{N}_0, \Delta t > 0 \text{ with } \mu\Delta t \leq T.$$

For the concrete determination of $\hat{C}(\lambda, \Delta t)$ from (6.31) we consider a term $C_j = a_j E_j$ from $C(\lambda, \Delta t) = \sum_j C_j$ (cf. (6.22) and (6.19)):

$$(C_j U)_\nu := a_j U_{\nu+j} \quad \text{for } U \in \ell^2.$$

As stated above, $\hat{C}_j \hat{U} = \mathcal{F}^{-1} C_j U$ holds with $\hat{U} = \mathcal{F}^{-1} U$; i.e., $\hat{U} \in L^2(0, 2\pi)$ is the Fourier series $\frac{1}{\sqrt{2\pi}} \sum_{\alpha\in\mathbb{Z}} U_\alpha e^{i\alpha\xi}$ with the coefficients $U = (U_\alpha)_{\alpha\in\mathbb{Z}} \in \ell^2$. Fourier synthesis of $C_j U = (a_j U_{\alpha+j})_{\alpha\in\mathbb{Z}}$ yields

$$\mathcal{F}^{-1} C_j U = \frac{1}{\sqrt{2\pi}} \sum_{\alpha\in\mathbb{Z}} (a_j U_{\alpha+j}) e^{i\alpha\xi} = a_j \frac{1}{\sqrt{2\pi}} \sum_{\alpha\in\mathbb{Z}} U_{\alpha+j} e^{i\alpha\xi}$$

$$\underset{\beta=\alpha+j}{=} a_j \frac{1}{\sqrt{2\pi}} \sum_{\beta\in\mathbb{Z}} U_\beta e^{i(\beta-j)\xi} = a_j e^{-ij\xi} \frac{1}{\sqrt{2\pi}} \sum_{\beta\in\mathbb{Z}} U_\beta e^{i\beta\xi} = a_j e^{-ij\xi} \hat{U}.$$

A comparison with $\mathcal{F}^{-1} C_j U = \hat{C}_j \hat{U}$ shows that $\hat{C}_j = a_j e^{-ij\xi}$; i.e., the linear mapping $\hat{C}_j : L^2(0, 2\pi) \to L^2(0, 2\pi)$ is the multiplication by the function $a_j e^{-ij\xi}$. Since $\hat{C}(\lambda, \Delta t) = \mathcal{F}^{-1} C(\lambda, \Delta t)\mathcal{F} = \mathcal{F}^{-1} \sum_j C_j \mathcal{F} = \sum_j \mathcal{F}^{-1} C_j \mathcal{F} = \sum_j \hat{C}_j$, we obtain the following remark.

Remark 6.42. *Consider the difference operator $C(\lambda, \Delta t) = \sum_{j\in\mathbb{Z}} a_j E_j$. The Fourier transformed operator is $\hat{C}(\lambda, \Delta t)=\sum_{j\in\mathbb{Z}} a_j e^{-ij\xi}$. Application of $\hat{C}(\lambda, \Delta t)$ corresponds to the multiplication by the trigonometric polynomial* [12]

$$G(\xi) := \sum_{j\in\mathbb{Z}} a_j e^{-ij\xi}, \tag{6.34}$$

which is called the 'characteristic function' or 'symbol' of $C(\lambda, \Delta t)$.

Exercise 6.43. *Let $\phi \in C(\mathbb{R})$ be a bounded continuous function. Define the multiplication operator $\Phi : B \to B$ by $(\Phi(f))(\xi) := \phi(\xi)f(\xi)$ for all $\xi \in \mathbb{R}$. Prove for both cases $B = C(\mathbb{R})$ and $B = L^2(\mathbb{R})$ that $\Phi \in \mathcal{L}(B, B)$ possesses the operator norm $\|\Phi\|_{B\leftarrow B} = \|\phi\|_\infty$.*

Therefore, the identity (6.33) becomes

$$\|C(\lambda, \Delta t)^\mu\|_{\ell^2\leftarrow\ell^2} = \sup\{|G(\xi)^\mu| : \xi \in \mathbb{R}\}. \tag{6.35}$$

Here, $\xi \in \mathbb{R}$ can be replaced by $\xi \in [0, 2\pi)$, since G is 2π-periodic. [13]

---

[12] The parameters $\lambda, \Delta t$ are omitted in $G(\xi) = G(\xi; \lambda, \Delta t)$. Note that a_j depends on $\lambda, \Delta t$.

[13] If the coefficients a_j are matrices (cf. Remark 6.18), $|G(\xi)^\mu|$ is to be replaced by the spectral norm $\|G(\xi)^\mu\|_2$.

6.5.3 Further Criteria

Application of the last exercise to $\hat{C}(\lambda, \Delta t)^\mu = \left(\sum_{j\in\mathbb{Z}} a_j e^{-ij\xi}\right)^\mu$ yields

$$\|\hat{C}(\lambda, \Delta t)^\mu\|_{L^2(0,2\pi)\leftarrow L^2(0,2\pi)} = \left\|\left(\sum_{j\in\mathbb{Z}} a_j e^{-ij\xi}\right)^\mu\right\|_\infty$$

$$= \left\|\sum_{j\in\mathbb{Z}} a_j e^{-ij\xi}\right\|_\infty^\mu. \qquad (6.36)$$

Note that (6.36) is only valid for *scalar* coefficients a_j. Together with Remark 6.41, (6.36) leads to the following theorem.

Theorem 6.44. *The difference scheme (6.19) is ℓ^2 stable if and only if the characteristic function $G(\xi) = \sum_{j\in\mathbb{Z}} a_j e^{-ij\xi}$ satisfies the estimate (6.37) with suitable K_λ:*

$$|G(\xi)| \le 1 + K_\lambda \Delta t \qquad \text{for all } \xi \in \mathbb{R}. \qquad (6.37)$$

Proof. (i) (6.37) implies $\|C(\lambda, \Delta t)\|_{\ell^2\leftarrow\ell^2} = \|\hat{C}(\lambda, \Delta t)\|_{L^2(0,2\pi)\leftarrow L^2(0,2\pi)} = \|G\|_\infty \le 1 + K_\lambda \Delta t$, so that Corollary 6.28 proves stability.

(ii) Set $c(\Delta t) := (\|G\|_\infty - 1)/\Delta t$. If there is no constant K_λ with (6.37), $c(\Delta t) \to \infty$ follows for $\Delta t \to 0$. Exercise 6.32a and (6.36) imply instability. $\square$

Proof of Criterion 6.31 in the case of ℓ^2. Since inequality $\sum a_j \ge 1 + \Delta t\, c(\Delta t)$ with $\lim_{\Delta t\to\infty} c(\Delta t) = \infty$ is supposed and $\|G\|_\infty \ge |G(0)| = |\sum a_j|$ holds, Exercise 6.32a proves ℓ^2 instability. $\square$

The previous analysis refers to ℓ^2. However, the characteristic function has also consequences for ℓ^∞.

Lemma 6.45. *Let $C(\lambda, \Delta t)$ be of the form (6.19) with constant coefficients a_j. Then ℓ^∞ stability implies ℓ^2 stability:*

$$\|C(\lambda, \Delta t)^\mu\|_{\ell^\infty\leftarrow\ell^\infty} \ge \|G^\mu\|_\infty = \|C(\lambda, \Delta t)^\mu\|_{\ell^2\leftarrow\ell^2} \quad \text{for all } \mu \in \mathbb{N}_0.$$

The negation of this statement is: if the difference scheme (6.19) is ℓ^2 unstable, it is also ℓ^∞ unstable.

Proof. Choose the special initial value U^0 with $U_\nu^0 = e^{i\nu\xi}$, where $\xi \in \mathbb{R}$ is characterised by $|G(\xi)| = \|G\|_\infty$. Note that $U^0 \in \ell^\infty$ and $\|U^0\|_\infty = 1$. Application of $C(\lambda, \Delta t)$ yields $U^1 = C(\lambda, \Delta t)U^0$ with

$$U_\nu^1 = \sum_{j\in\mathbb{Z}} a_j U_{\nu+j}^0 = \sum_{j\in\mathbb{Z}} a_j e^{i(\nu+j)\xi} = e^{i\nu\xi}\sum_{j\in\mathbb{Z}} a_j e^{ij\xi} = G(\xi)U_\nu^0,$$

so that $C(\lambda, \Delta t)^\mu U^0 = G(\xi)^\mu U^0$ and $\|C(\lambda, \Delta t)^\mu U^0\|_\infty = \|G(\xi)^\mu U^0\|_\infty = |G(\xi)|^\mu \|U^0\|_\infty = \|G^\mu\|_\infty \|U^0\|_\infty$. This proves the assertion. $\square$

Below, the examples from §6.3.3 are again analysed with regard to (in)stability.

Example 6.46. *(a)* ***Scheme (6.17a)*** *with* $a_0 = 1 - a\lambda$, $a_1 = a\lambda$. *For* λ *satisfying* $0 \leq a\lambda \leq 1$, *stability is already confirmed in Example 6.33a. The associated characteristic function*

$$G(\xi) := 1 - a\lambda + a\lambda e^{-i\xi} = 1 - a\lambda(1 - \cos\xi) - ia\lambda\sin\xi$$

has modulus $|G(\pi)| = |1 - 2a\lambda|$ *at* $\xi = \pi$. *Since* $|G(\pi)| > 1$ *for all* $a\lambda \notin [0, 1]$, *Theorem 6.44 and Lemma 6.45 prove instability with respect to* ℓ^2 *and* ℓ^∞ *for* $a\lambda$ *outside of* $[0, 1]$. *Hence, the scheme (6.17a) is conditionally stable (with respect to* ℓ^2 *and* ℓ^∞) *under the restriction* $a\lambda \in [0, 1]$, *and unstable otherwise.*

(b) ***Scheme (6.17b)*** *with* $a_{-1} = -\frac{a\lambda}{2}$, $a_0 = 1$, $a_1 = \frac{a\lambda}{2}$. *The associated characteristic function*

$$G(\xi) := -\frac{a\lambda}{2}e^{i\xi} + 1 + \frac{a\lambda}{2}e^{-i\xi} = 1 - ia\lambda\sin\xi$$

has maximum norm $\|G\|_\infty = \sqrt{1 + |a\lambda|^2}$ *and, therefore, except the trivial case* $a = 0$, *the scheme is always unstable (with respect to* ℓ^2 *and* ℓ^∞).

(c) ***Scheme (6.17c)*** *with* $a_{-1} = \frac{1-a\lambda}{2}$, $a_1 = \frac{1+a\lambda}{2}$. *If* $|a\lambda| \leq 1$, *Example 6.33c shows stability. Because of*

$$G(\xi) := \frac{1-a\lambda}{2}e^{i\xi} + \frac{1+a\lambda}{2}e^{-i\xi} \quad \text{and} \quad |G(\xi)|^2 = \cos^2\xi + |a\lambda|^2\sin^2\xi,$$

the bound $\|G\|_\infty = \max\{1, |a\lambda|^2\}$ *follows. Hence, the scheme is conditionally stable for* $|a\lambda| \leq 1$, *but unstable for* $|a\lambda| > 1$ *(with respect to* ℓ^2 *and* ℓ^∞).

(d) ***Scheme (6.18)*** *with* $a_{-1} = \lambda$, $a_0 = 1 - 2\lambda$, $a_1 = \lambda$. *For* $\lambda \in (0, 1/2]$, *stability is shown in Example 6.33d. Because of*

$$G(\xi) := \lambda e^{i\xi} + 1 - 2\lambda + \lambda e^{-i\xi} = 1 - 2\lambda(1 - \cos\xi)$$

and $\|G\|_\infty = |G(\pi)| = |1 - 4\lambda|$ *the scheme is conditionally stable for* $\lambda \in (0, 1/2]$, *but unstable for* $\lambda > 1/2$ *(with respect to* ℓ^2 *and* ℓ^∞).

Lemma 6.45 yields a direct relation between G and ℓ^2 stability. Because of the inequality $\|C(\lambda, \Delta t)^\mu\|_{\ell^\infty \leftarrow \ell^\infty} \geq \|C(\lambda, \Delta t)^\mu\|_{\ell^2 \leftarrow \ell^2}$, boundedness of powers of G is necessary for ℓ^∞ stability. As we shall see, this condition is not sufficient for ℓ^∞ stability. A complete characterisation of ℓ^∞ stability is given by Thomée [24].

Theorem 6.47. *Assume a difference scheme (6.16) with* $G(\xi) = \sum_{j\in\mathbb{Z}} a_j e^{-ij\xi}$ *(cf. (6.34)). Then the scheme is* ℓ^∞ *stable if and only if either condition (a) or (b) is fulfilled:*

(a) $G(\xi) = ce^{-ij\xi}$ *with* $|c| = 1$,

(b) the set $\{\xi \in [-\pi, \pi] : |G(\xi)| = 1\} = \{\xi_1, \ldots, \xi_N\}$ *has finite cardinality, and there are numbers* $\alpha_k \in \mathbb{R}$, $\beta_k \in \mathbb{N}$, *and* $\gamma_k \in \mathbb{C}$ *with* $\Re\gamma_k > 0$, *such that*

$$G(\xi_k + \xi) = G(\xi_k)\exp\left(i\alpha_k\xi - \gamma_k\xi^{2\beta_k}(1 + o(1))\right) \quad \text{as } \xi \to 0 \quad (1 \leq k \leq N).$$

The CFL condition is a necessary condition for convergence. 'CFL' abbreviates the names of the authors Courant, Friedrichs, Lewy of [1].[14]

Criterion 6.48 (Courant–Friedrichs–Lewy). *Let the hyperbolic differential equation $\frac{\partial}{\partial t} u = a \frac{\partial}{\partial x} u$ be discretised by the explicit difference scheme (6.16). Set*

$$J_1 = \min\{j \in \mathbb{Z} : a_j \neq 0\},$$
$$J_2 = \max\{j \in \mathbb{Z} : a_j \neq 0\},$$

so that the sum $\sum\limits_{j \in \mathbb{Z}} a_j U^\mu_{\nu+j}$ can be reduced to $\sum\limits_{J_1 \leq j \leq J_2} a_j U^\mu_{\nu+j}$. If

$$a\lambda \notin [J_1, J_2],$$

the scheme is not convergent (with respect to any norm). Therefore, $a\lambda \in [J_1, J_2]$ is a necessary convergence condition.

Because of the equivalence theorem, stability cannot hold for $a\lambda \notin [J_1, J_2]$. Hence, the CFL criterion is also a *stability* criterion.

Proof of Criterion 6.48. For an arbitrary space-time point (t, x) with $t > 0$, the solution is $u(t, x) = u_0(x + at)$ according to Lemma 6.2. One verifies that a grid point $(t, x) = (\mu \Delta t, \nu \Delta x)$ of the solution U^μ_ν depends only on the initial values U^0_k with $k \in [\nu + \mu J_1, \nu + \mu J_2]$. Multiplication by Δx shows

$$\begin{aligned}
x_k = k\Delta x &\in [x + \mu \Delta x J_1, x + \mu \Delta x J_2] \\
&= [x + \mu \Delta x J_1, x + \mu \Delta x J_2] \\
&= [x + t J_1/\lambda, x + t J_2/\lambda].
\end{aligned}$$

If $a\lambda \notin [J_1, J_2]$, then $x + at \notin [x + t J_1/\lambda, x + t J_2/\lambda]$. Therefore, the computation of U^μ_ν does not use those data, on which the solution $u(t, x)$ depends. To be more precise, choose an initial value $u_0 \in C^\infty_0(\mathbb{R})$ with

$$u_0(a) \neq 0, \qquad \mathrm{supp}(u_0) \subset \mathbb{R}\backslash[J_1/\lambda, J_2/\lambda].$$

Then $U(1, x) = 0$ holds in a neighbourhood of $x = 0$, while the true solution satisfies $u(1, 0) = u_0(a) \neq 0$ and even $u(1, x) \neq 0$ in a neighbourhood of $x = 0$. As a consequence, U^μ_ν cannot converge to u. $\square$

As an example, take the scheme (6.17a). Since $a_0 = 1 - a\lambda$ and $a_1 = a\lambda$, the CFL bounds are $J_1 = 0$ and $J_2 = 1$. The necessary CFL condition $a\lambda \in [0, 1]$ coincides for Example 6.46a with the exact stability property.

We summarise the mentioned necessary and/or sufficient conditions in the following table.

---

[14] The original paper is written in German. An English translation can be found in [3, Appendix C].

statement	name or comment	necessary ℓ^2	ℓ^∞	sufficient ℓ^2	ℓ^∞
Criterion 6.26				*	*
Corollary 6.28				*	*
Criterion 6.30	positive scheme			*	*
Criterion 6.31		*	*		
Lemma 6.34	perturbation			*	*
Criterion 6.36		*	*		
Criterion 6.38	almost normal	*		*	
Theorem 6.44		*		*	
Lemma 6.45			*	*	
Theorem 6.47			*		*
Criterion 6.48	CFL	*	*		
Criterion 6.52		*		*	
Criterion 6.54(a)	von Neumann	*	*		
Criterion 6.54(b)	von Neumann	*		*	
Lemma 6.55	Lax-Wendroff			*	
Criterion 6.57	Friedrichs			*	
Theorem 6.58	Friedrichs				*

6.5.4 Implicit Schemes

All schemes examined so far are, at best, conditionally stable. In order to obtain *unconditionally stable* schemes, one must admit implicit difference schemes.[15]

In the parabolic case $\frac{\partial u}{\partial t} = \frac{\partial^2 u}{\partial x^2}$, the second x-difference can be formed at time level $t + \Delta t$, i.e.,

$$\frac{\partial^2 u}{\partial x^2} \approx \frac{u(t + \Delta t, x - \Delta x) - 2u(t + \Delta t, x) + u(t + \Delta t, x + \Delta x)}{\Delta x^2},$$

while the time derivative $\frac{\partial}{\partial t} u$ becomes $\frac{u(t+\Delta t,x)-u(t,x)}{\Delta t}$ as before. This leads to

$$-\lambda U_{\nu-1}^{\mu+1} + (1 + 2\lambda) U_\nu^{\mu+1} - \lambda U_{\nu+1}^{\mu+1} = U_\nu^\mu. \tag{6.38}$$

Instead of the explicit form $U^{\mu+1} = C(\lambda, \Delta t)U^\mu$, one now obtains an implicit scheme of the form

$$C_1(\lambda, \Delta t)U^{\mu+1} = C_2(\lambda, \Delta t)U^\mu, \tag{6.39}$$

where in the present case the operators are given by

---

[15] The CFL criterion does not apply to implicit schemes, since formally implicit schemes can be viewed as explicit ones with an infinite sum (i.e., $J_1 = -\infty$, $J_2 = \infty$). Then $a\lambda \in [J_1, J_2] = \mathbb{R}$ is valid, and the CFL condition is always satisfied.

$$(C_1(\lambda, \Delta t)U)_\nu = \sum_{-1 \le j \le 1} a_{1,j} U_{\nu+j} \quad \text{with } a_{1,-1} = a_{1,1} = -\lambda, a_{1,0} = 1 + 2\lambda \quad \text{and}$$

$$C_2(\lambda, \Delta t)U = \sum_{j \in \mathbb{Z}} a_{2,j} U_{\nu+j} \quad \text{with} \quad a_{2,0} = 1, \quad \begin{cases} \text{all other coefficients} = 0; \\ \text{i.e., } C_2 = identity. \end{cases}$$

One would like to solve (6.39) with respect to $U^{\mu+1}$:

$$U^{\mu+1} = C(\lambda, \Delta t)U^\mu \quad \text{with } C(\lambda, \Delta t) := [C_1(\lambda, \Delta t)]^{-1} C_2(\lambda, \Delta t). \quad (6.40)$$

Therefore, the existence of the inverse $[C_1(\lambda, \Delta t)]^{-1}$ is to be investigated.

Lemma 6.49. (a) If the coefficients $a_{1,j}$ of $C_1(\lambda, \Delta t)$ satisfy the inequality

$$\left(\sum_{j \in \mathbb{Z} \setminus \{0\}} |a_{1,j}| \right) / |a_{1,0}| \le 1 - \varepsilon < 1 \quad \text{for some } \varepsilon > 0,$$

then the inverse exists and satisfies $\| [C_1(\lambda, \Delta t)]^{-1} \|_{\ell^p \leftarrow \ell^p} \le \frac{1}{\varepsilon |a_{1,0}|}$.
(b) The inverse $[C_1(\lambda, \Delta t)]^{-1}$ exists in $\mathcal{L}(\ell^2, \ell^2)$ if and only if the characteristic function $G_1(\xi)$ of C_1 satisfies an inequality $|G_1(\xi)| \ge \eta > 0$. The norm equals

$$\| [C_1(\lambda, \Delta t)]^{-1} \|_{\ell^2 \leftarrow \ell^2} = 1 / \inf_{\xi \in \mathbb{R}} |G_1(\xi)|.$$

Proof. (i) The equation $C_1 V = U$ is equivalent to the fixed-point equation

$$V = \frac{1}{a_{1,0}} (U - C_1' V) =: \Phi(V) \quad \text{with } C_1' := C_1 - a_{1,0} I.$$

The contraction constant of Φ is $\|C_1'\|_{\ell^p \leftarrow \ell^p} / |a_{1,0}|$. By Remark 6.27,

$$\frac{\|C_1'\|_{\ell^p \leftarrow \ell^p}}{|a_{1,0}|} \le \frac{\sum_{j \in \mathbb{Z} \setminus \{0\}} |a_{1,j}|}{|a_{1,0}|} \le 1 - \varepsilon$$

is valid, so that a unique inverse exists. The estimate

$$\|V\|_{\ell^p} \le \frac{1}{|a_{1,0}|} \|U\|_{\ell^p} + (1 - \varepsilon) \|V\|_{\ell^p}$$

shows that $\varepsilon \|V\|_{\ell^p} \le \frac{1}{|a_{1,0}|} \|U\|_{\ell^p}$, so that $\| [C_1(\lambda, \Delta t)]^{-1} \|_{\ell^p \leftarrow \ell^p}$ has the bound $1/ (\varepsilon |a_{1,0}|)$.

(ii) $\hat{C}_1(\lambda, \Delta t) = \mathcal{F}^{-1} C_1(\lambda, \Delta t) \mathcal{F}$ is the Fourier transformation of C_1. $\hat{C}_1(\lambda, \Delta t)$ is the operator $\hat{U} \mapsto G_1 \hat{U}$. Obviously, the multiplication operator $M \in \mathcal{L}(\ell^2, \ell^2)$ with $M\hat{U} := (1/G_1) \hat{U}$ is the inverse of $\hat{C}_1$. The norm of M is

$$\|M\|_{\ell^2 \leftarrow \ell^2} = \|\hat{C}_1^{-1}\|_{\ell^2 \leftarrow \ell^2} = \|1/G_1\|_\infty \le 1/\eta.$$

On the other hand, one verifies that $\|\hat{C}_1^{-1}\|_{\ell^2 \leftarrow \ell^2}$ cannot be finite if $\inf |G_1|$ equals zero. Since the Fourier transform does not change the ℓ^2 norm, the equality $\|C_1^{-1}\|_{\ell^2 \leftarrow \ell^2} = \|\hat{C}_1^{-1}\|_{\ell^2 \leftarrow \ell^2} = \|1/G_1\|_\infty = 1/\inf_{\xi \in \mathbb{R}} |G_1(\xi)|$ follows. $\quad\square$

Example 6.50. *The scheme (6.38) is ℓ^2 stable for all $\lambda = \Delta t/\Delta x^2$; i.e., it is uncon-ditionally stable.*

Proof. $G_1(\xi) = -\lambda e^{i\xi} + 1 + 2\lambda - \lambda e^{-i\xi} = 1 + 2\lambda(1 - \cos x) \geq \inf_{\xi \in \mathbb{R}} |G_1(\xi)| = 1$ is the characteristic function of C_1. Since $C_2 = I$, we obtain

$$C(\lambda, \Delta t) := [C_1(\lambda, \Delta t)]^{-1} C_2(\lambda, \Delta t) = [C_1(\lambda, \Delta t)]^{-1}$$

and $\|C_1^{-1}\|_{\ell^2 \leftarrow \ell^2} = 1/\inf_{\xi \in \mathbb{R}} |G_1(\xi)| = 1$. Stability follows by Criterion 6.26. $\quad\square$

Remark 6.51. $[C_1(\lambda, \Delta t)]^{-1}$ *from above is equal to the infinite operator*

$$[C_1(\lambda, \Delta t)]^{-1} = \sum_{j \in \mathbb{Z}} a_j(\lambda) E_j \quad \text{with } a_j = \frac{1}{\sqrt{1 + 4\lambda}} \left(\frac{2\lambda}{2\lambda + 1 + \sqrt{1 + 4\lambda}} \right)^{|j|};$$

i.e., the implicit scheme (6.38) is identical to $U_\nu^{\mu+1} = \sum_{j \in \mathbb{Z}} a_j U_{\nu+j}^\mu$.

Proof. Check that $C_1(\lambda, \Delta t) \sum_{j \in \mathbb{Z}} a_j E_j = identity$. $\quad\square$

The general case of an implicit scheme (6.39) is treated in the next criterion (its proof is identical to that of Theorem 4.44).

Criterion 6.52. *The scheme (6.39) is ℓ^2 stable if and only if the characteristic func-tion*

$$G(\xi) := G_2(\xi)/G_1(\xi)$$

satisfies condition (6.37).

Example 6.53. *A modification of (6.38) is the so-called* theta scheme

$$- \lambda \Theta U_{\nu-1}^{\mu+1} + (1 + 2\lambda \Theta) U_\nu^{\mu+1} - \lambda \Theta U_{\nu+1}^{\mu+1} \tag{6.41}$$
$$= \lambda(1 - \Theta) U_{\nu-1}^\mu + (1 - 2\lambda(1 - \Theta)) U_\nu^\mu + \lambda(1 - \Theta) U_{\nu+1}^\mu$$

for $\Theta \in [0, 1]$. For $\Theta = 0$ and $\Theta = 1$ we regain (6.18) and (6.38), respectively. For $\Theta = 1/2$, (6.41) is called[16] the Crank–Nicolson *scheme. The scheme (6.41) is unconditionally ℓ^2 stable for $\Theta \in [1/2, 1]$, whereas in the case of $\Theta \in [0, 1/2)$ it is conditionally ℓ^2 stable for*

$$\lambda \leq 1/(2(1 - 2\Theta)).$$

In all stable cases, $\|C(\lambda, \Delta t)\|_{\ell^2 \leftarrow \ell^2} = 1$ holds.

---

[16] Occasionally, one finds the incorrect spelling 'Crank–Nicholson'. The second author of [2] is Mrs. Phyllis Nicolson (1917–1968).

Proof. Let $(DU)_\nu = -U_{\nu-1} + 2U_\nu - U_{\nu+1}$ be the negative second difference operator. The characteristic function of D is $G_D(\xi) = 2 - 2\cos\xi = 4\sin^2(\xi/2)$. The operators C_1, C_2 from (6.39) in the case of (6.41) are

$$C_1(\lambda, \Delta t) = I + \lambda\Theta D \quad\text{and}\quad C_2(\lambda, \Delta t) = I - \lambda(1 - \Theta)D.$$

The associated functions are

$$G_1(\xi) = 1 + \lambda\Theta G_D(\xi) \quad\text{and}\quad G_2(\xi) = 1 - \lambda(1 - \Theta)G_D(\xi),$$

so that

$$G(\xi) = \frac{1 - \lambda(1 - \Theta)G_D(\xi)}{1 + \lambda\Theta G_D(\xi)}.$$

The function $\frac{1-\lambda(1-\Theta)X}{1+\lambda\Theta X}$ is monotonically decreasing with respect to X, so that the maximum of $|G(\xi)|$ is taken at $X = 0 = G_D(0)$ or $X = 4 = G_D(\pi)$:

$$\|G\|_\infty = \max\{G(0), -G(\pi)\} = \max\left\{1, \frac{4\lambda(1 - \Theta) - 1}{1 + 4\lambda\Theta}\right\}.$$

If $\Theta \in [\frac{1}{2}, 1]$, then

$$-G(\pi) = \frac{4\lambda(1 - \Theta) - 1}{1 + 4\lambda\Theta} = \frac{4\lambda}{1 + 4\lambda\Theta} - 1 \underset{\Theta \geq \frac{1}{2}}{\leq} \frac{4\lambda}{1 + 2\lambda} - 1 = \frac{2\lambda - 1}{2\lambda + 1} \leq 1$$

proves $\|G\|_\infty = 1$.

In the case of $\Theta \in [0, 1/2)$, the choice of λ must ensure the estimate $-G(\pi) \leq 1$. Equivalent statements are

$$4\lambda(1 - \Theta) - 1 \leq 1 + 4\lambda\Theta \Leftrightarrow 4\lambda(1 - 2\Theta) \leq 2 \Leftrightarrow \lambda \leq 1/(2(1 - 2\Theta)). \quad\square$$

6.5.5 Vector-Valued Grid Functions

So far, ℓ^p has been the set the complex-valued sequences $(U_\nu)_{\nu\in\mathbb{Z}}$, $U_\nu \in \mathbb{C}$. If the equation $\frac{\partial}{\partial t}u(t) = Au(t)$ from (6.1a) is vector-valued (values in $\mathbb{C}^N$), also the grid functions $(U_\nu)_{\nu\in\mathbb{Z}}$ must be vector-valued:

$$\ell^p = \{U = (U_\nu)_{\nu\in\mathbb{Z}} : U_\nu \in \mathbb{C}^N\} \quad\text{with the norms}$$

$$\|U\|_2 = \sqrt{\sum_{\nu\in\mathbb{Z}} \|U_\nu\|_2^2}, \quad \|U\|_\infty = \sup_{\nu\in\mathbb{Z}} \|U_\nu\|_\infty,$$

where $\|U_\nu\|_p$ is the Euclidean norm in $\mathbb{C}^N$ ($p = 2$) or the maximum norm in $\mathbb{C}^N$ ($p = \infty$, cf. Remark 6.18).

Now, the coefficients a_j in the difference scheme (6.19) are $N \times N$ matrices. The statements concerning consistency, convergence, and stability remain unchanged (only the norms are to be interpreted differently). However, the criteria (see §6.5 and later) are to be generalised to the case $N > 1$.

Criterion 6.26 remains valid without any change.

The estimate in Remark 6.27 becomes

$$\|C(\lambda, \Delta t)\|_{\ell^p \leftarrow \ell^p} \leq \sum_j \|a_j\|_p,$$

where $\|.\|_2$ is the spectral norm and $\|.\|_\infty$ is the row-sum norm for $N \times N$ matrices.

In Corollary 6.28 one has to replace $\sum |a_j|$ by $\sum_j \|a_j\|_p$.

When we form $G(\xi) = \sum_{j \in \mathbb{Z}} a_j e^{-ij\xi}$ by the Fourier transform, $G(\xi)$ is now an $N \times N$ matrix-valued function. The characterisation (6.35) becomes

$$\|C(\lambda, \Delta t)^\mu\|_{\ell^2 \leftarrow \ell^2} = \sup\{\|G(\xi)^\mu\|_2 : \xi \in [0, 2\pi)\}. \tag{6.42}$$

Instead of the relatively abstract operator $C(\lambda, \Delta t)$, one has now to investigate the boundedness of the $N \times N$ matrices $G(\xi)^\mu$.

According to Criterion 6.36, $\rho(C(\lambda, \Delta t)) \leq 1 + \mathcal{O}(\Delta t)$ is a necessary condition. Since the unitary Fourier transform does not change the spectra, $C(\lambda, \Delta t)$ and $\{G(\xi) : \xi \in [0, 2\pi)\}$ have identical eigenvalues. This will lead to the von Neumann condition. Here the *spectral radius* $\rho(G(\xi))$ is defined in (6.25). Part (b) corresponds to Remark 6.37.

Criterion 6.54 (von Neumann condition). *(a) A necessary condition for stability (with respect to ℓ^2 and ℓ^∞) is*

$$\sup\{\rho(G(\xi)) : \xi \in [0, 2\pi)\} \leq 1 + \mathcal{O}(\Delta t).$$

(b) If all matrices $G(\xi)$ are normal, this condition is even sufficient for ℓ^2 stability.

The following statement uses the *numerical radius* of a matrix:

$$r(A) := \sup_{0 \neq v \in \mathbb{C}^N} \left| \frac{\langle Av, v \rangle}{\|v\|_2^2} \right|.$$

Lemma 6.55 (Lax–Wendroff condition, [18]). *Suppose that there is a constant K_{LW} such that*

$$r(G(\xi)) \leq 1 + K_{LW} \Delta t \qquad \text{for all } \xi \in [0, 1].$$

Then the scheme is ℓ^2 stable.

Proof. The numerical radius of general square matrices A possesses the properties

$$\|A\|_2 \le 2r(A), \qquad r(A^n) \le r(A)^n \quad \text{for } n \in \mathbb{N}_0$$

(cf. §2.9.5 in [6] or [7]). Hence,

$$\|G(\xi)^n\|_2 \le 2r(G(\xi)^n) \le 2r(G(\xi))^n \le 2\,(1 + K_{LW}\Delta t)^n \le 2\exp(K_{LW}T)$$

holds for all n with $n\Delta t \le T$ and all $\xi \in [0, 2\pi)$. Because of (6.42), the assertion is proved. $\square$

Definition 6.29 carries over to the matrix case when we replace non-negative reals by positive semidefinite matrices; i.e., a *positive difference scheme* is characterised by 'a_j positive semidefinite'.

Exercise 6.56. *Suppose that the positive semidefinite coefficients a_j are either*

(i)   all diagonal or
(ii)   simultaneously diagonalisable; i.e., there is a transformation S such that the matrices $d_j = S a_j S^{-1}$ are diagonal for all j.

Show that analogous to Criterion 6.30, the difference scheme (6.19) is ℓ^2 and ℓ^∞ stable if $\sum a_j = I$ (cf. (6.24)).

However, also without simultaneous diagonalisability, the criterion can be generalised. Note that even x-dependent coefficients $a_j = a_j(x)$ are allowed.

Criterion 6.57 (Friedrichs [5]). *Suppose that the difference scheme (6.19) has positive semidefinite coefficients a_j satisfying the consistency condition $\sum_{j\in\mathbb{Z}} a_j = I$ (cf. (6.24)). The coefficients a_j must be either constant or the following three conditions must hold:*
(i)   the hyperbolic case with $\lambda = \Delta t/\Delta x$ is given (cf. (6.11)),
(ii) $a_j(\cdot)$ are globally Lipschitz continuous in $\mathbb{R}$ with Lipschitz constant L_j,
(iii) $B := \sum_{j\in\mathbb{Z}} j L_j < \infty$.
Then

$$\|C(\lambda, \Delta t)\|_{\ell^2 \leftarrow \ell^2} \le 1 + C_L \Delta t$$

holds with $C_L = B/(2\lambda)$ implying ℓ^2 stability.

Proof. We prove the general case (constant coefficients correspond to $L_j = 0$). In

$$(C(\lambda, \Delta t)V, U)_{\ell^2} = \Delta x \sum_{\nu\in\mathbb{Z}} \sum_{j\in\mathbb{Z}} \langle a_j(\nu\Delta x)V_{\nu+j}, U_\nu\rangle \qquad (U, V \in \ell^2) \quad (6.43)$$

$\langle\cdot, \cdot\rangle$ denotes the scalar product of $\mathbb{C}^N$ (N is the dimension of $U_\nu, V_\nu \in \mathbb{C}^N$).

Any positive semidefinite matrix M fulfils $|\langle Mx, y\rangle| \le \frac{1}{2}\langle Mx, x\rangle + \frac{1}{2}\langle My, y\rangle$ for $x, y \in \mathbb{C}^N$. Application to (6.43) yields

$$\left| \Delta x \sum_{\nu \in \mathbb{Z}} \sum_{j \in \mathbb{Z}} \langle a_j(\nu \Delta x) V_{\nu+j}, U_\nu \rangle \right|$$

$$\leq \frac{\Delta x}{2} \sum_{\nu \in \mathbb{Z}} \sum_{j \in \mathbb{Z}} \langle a_j(\nu \Delta x) U_\nu, U_\nu \rangle + \frac{\Delta x}{2} \sum_{\nu \in \mathbb{Z}} \sum_{j \in \mathbb{Z}} \langle a_j(\nu \Delta x) V_{\nu+j}, V_{\nu+j} \rangle .$$

The first term is $\frac{\Delta x}{2} \sum_{\nu \in \mathbb{Z}} \left\langle \sum_{j \in \mathbb{Z}} a_j(\nu \Delta x) U_\nu, U_\nu \right\rangle \underset{(6.24)}{=} \frac{\Delta x}{2} \sum_{\nu \in \mathbb{Z}} \|U_\nu\|_2^2 = \frac{1}{2} \|U\|_{\ell^2}^2$.

Substitute ν in the second term by $\mu = \nu + j$:

$$\frac{\Delta x}{2} \sum_{\nu \in \mathbb{Z}} \sum_{j \in \mathbb{Z}} \langle a_j(\nu \Delta x) V_{\nu+j}, V_{\nu+j} \rangle = \frac{\Delta x}{2} \sum_{\mu \in \mathbb{Z}} \sum_{j \in \mathbb{Z}} \langle a_j((\mu - j)\,\Delta x) V_\mu, V_\mu \rangle .$$

Because of Lipschitz continuity, $\|a_j((\mu - j)\,\Delta x) - a_j(\mu \Delta x)\|_2 \leq L_j j \Delta x$ holds.
With $B = \sum_{j \in \mathbb{Z}} j L_j < \infty$ (assumption (iii)) we estimate by

$$\frac{\Delta x}{2} \sum_{\mu \in \mathbb{Z}} \sum_{j \in \mathbb{Z}} \langle a_j((\mu - j)\,\Delta x) V_\mu, V_\mu \rangle$$

$$\underset{(6.24)}{=} \frac{\Delta x}{2} \sum_{\mu \in \mathbb{Z}} \sum_{j \in \mathbb{Z}} \langle [a_j((\mu - j)\,\Delta x) - a_j(\mu \Delta x)] V_\mu, V_\mu \rangle + \frac{\Delta x}{2} \|V\|_{\ell^2}^2$$

$$\leq \frac{1 + B \Delta x}{2} \|V\|_{\ell^2}^2 .$$

From the previous estimates we obtain $|(CV, U)_{\ell^2}| \leq \frac{1}{2} \|U\|_{\ell^2}^2 + \frac{1 + B \Delta x}{2} \|V\|_{\ell^2}^2$.
Characterising the norm by

$$\|C(\lambda, \Delta t)\|_{\ell^2 \leftarrow \ell^2} = \sup_{\|U\|_{\ell^2} = \|V\|_{\ell^2} = 1} |(CV, U)_{\ell^2}|,$$

we obtain the inequality $\|C(\lambda, \Delta t)\|_{\ell^2 \leftarrow \ell^2} \leq 1 + \frac{B \Delta x}{2} = 1 + C_L \Delta t$. The last step
uses either $\Delta x = \Delta t / \lambda$ from assumption (i) or $B = 0$ (constant coefficients). $\quad \square$

We add some comments concerning the latter criterion.

(1) Inequality $\|C(\lambda, \Delta t)\|_{\ell^2 \leftarrow \ell^2} \leq 1 + \mathcal{O}(\Delta x) = 1 + \mathcal{O}(\sqrt{\Delta t})$ is not sufficient for
stability in the parabolic case with $\lambda = \Delta t / \Delta x^2$.

(2) If the coefficients a_j are constant, the vanishing values $L_j = B = C_L = 0$ show
that $\|C(\lambda, \Delta t)\|_{\ell^2 \leftarrow \ell^2} = 1$.

(3) If only finitely many coefficients are different from zero, $B = \sum_{j \in \mathbb{Z}} j L_j < \infty$
is satisfied.

A positive difference scheme can be obtained as discretisation of the *symmetric
hyperbolic system of differential equations*

$$\frac{\partial}{\partial t} u + A(x) \frac{\partial}{\partial x} u = 0 \qquad (A(x) \text{ symmetric } N \times N \text{ matrix}, u \in \mathbb{C}^N). \qquad (6.44)$$

Replacing $\frac{\partial}{\partial x} u$ by $\frac{u(t,x+\Delta x)-u(t,x-\Delta x)}{2r\Delta x}$ and $\frac{\partial}{\partial t} u$ by $\frac{u(t+\Delta t,x)-\bar{u}(t,x)}{\Delta t}$ with the average $\bar{u}(t,x) = \frac{1}{2}[u(t,x+\Delta x) + u(t,x-\Delta x)]$, we are led to the difference scheme (*Friedrichs' scheme*)

$$(C(\lambda, \Delta t)U)_\nu = \frac{1}{2}\{[I - \lambda A(\nu\Delta x)]U_{\nu-1} + [I + \lambda A(\nu\Delta x)]U_{\nu+1}\}. \quad (6.45)$$

Choosing $0 < \lambda \le 1/\sup_{x\in\mathbb{R}}\|A(x)\|_2$, we obtain a positive scheme $C(\lambda, \Delta t)$. Hence ℓ^2 stability follows, provided that $A(x)$ is Lipschitz continuous.

Criterion 6.36 as well as Remark 6.37 and Criterion 6.38 remain valid in the vector-valued case.

Also in the vector-valued case, the Fourier transformed difference operator has the form $\hat{C}(\lambda, \Delta t)(\xi) = \sum_{j\in\mathbb{Z}} a_j e^{-ij\xi}$, where now $\hat{C}$ is a 2π-periodic function, whose values are $N \times N$ matrices. As in the scalar case, the stability estimate $\|C(\lambda, \Delta t)^\mu\|_{\ell^2\leftarrow\ell^2} \le K(\lambda)$ holds for all $\mu\Delta t \le T$ if and only if

$$\|\hat{C}(\lambda, \Delta t)^\mu\|_{L^2(0,2\pi)\leftarrow L^2(0,2\pi)} \le K(\lambda).$$

However, the equality $\|(\sum_{j\in\mathbb{Z}} a_j e^{-ij\xi})^\mu\|_\infty = \|\sum_{j\in\mathbb{Z}} a_j e^{-ij\xi}\|_\infty^\mu$ from (6.36) is, in general, no longer valid for the matrix-valued case, but becomes an inequality:

$$\|(\sum_{j\in\mathbb{Z}} a_j e^{-ij\xi})^\mu\|_\infty \le \|\sum_{j\in\mathbb{Z}} a_j e^{-ij\xi}\|_\infty^\mu.$$

The investigation of ℓ^∞ stability of a hyperbolic system is more involved. Following Mizohata [20], we define regular hyperbolicity. A system

$$\frac{\partial}{\partial t}u + A(t,x)\frac{\partial}{\partial x}u = 0 \quad (A \; N\times N \text{ matrix}, \; u \in \mathbb{C}^N, \; x \in \mathbb{R}, \; 0 \le t \le T) \quad (6.46)$$

is called *regularly hyperbolic*, if the eigenvalues $d_i(t,x)$ $(1 \le i \le N)$ of $A(t,x)$ are real and distinct; i.e., there is some $\delta > 0$ such that

$$|d_i(t,x) - d_j(t,x)| \ge \delta \quad \text{for all } i \ne j \quad (1 \le i,j \le N, x \in \mathbb{R}, 0 \le t \le T).$$

The following stability result is proved by Tomoeda [26].

Theorem 6.58. *Suppose that (6.46) is a regularly hyperbolic system satisfying the smoothness condition*

$$\sup\{|A(t,x)|, |A_t(t,x)|, |A_x(t,x)| : x \in \mathbb{R}, 0 \le t \le T\} < \infty.$$

Then Friedrichs' scheme (6.45) with

$$0 < \lambda < 1/\sup_{x\in\mathbb{R},0\le t\le T}|d_i(t,x)|$$

is ℓ^∞ stable.

6.5.6 Generalisations

In §6.5.5 we have already analysed the generalisation to the case of *systems* of differential equations with vector-valued solutions. Next, a series of further generalisations is discussed.

6.5.6.1 Spatially Multivariate Case

Instead of one spatial variables, we admit d variables $x = (x_1, \ldots, x_d)$ ($d = 3$ is the standard case). Then the heat equation (6.4) becomes $\frac{\partial u}{\partial t} = \Delta u$ for $t > 0$ with the Laplace operator $\Delta = \sum_{k=1}^d \frac{\partial^2}{\partial x_k^2}$. The representation (6.5) of u can be adapted to the d-dimensional situation.

The hyperbolic differential equation $\frac{\partial}{\partial t} u = a \frac{\partial}{\partial x} u$ is generalised to

$$\frac{\partial}{\partial t} u = \sum_{k=1}^d A_k \frac{\partial u}{\partial x_k}. \tag{6.47}$$

For the discretisation, the grid $G_{\Delta x} = \{x = \nu \Delta x : \nu \in \mathbb{Z}\}$ from (6.9) is to be replaced by the d-dimensional grid[17]

$$G_{\Delta x} = \{x = \nu \Delta x : \nu \in \mathbb{Z}^d\}$$

with multi-indices $\nu = (\nu_1, \ldots, \nu_d)$, $\nu_j \in \mathbb{Z}$. Now, the Banach space ℓ^p is $\mathbb{C}^{\mathbb{Z}^d}$ equipped with the norms

$$\|U\|_{\ell^\infty} = \sup\{|U_\nu| : \nu \in \mathbb{Z}^d\} \quad \text{and} \quad \|U\|_{\ell^2} = \sqrt{\Delta x^d \sum_{\nu \in \mathbb{Z}^d} |U_\nu|^2}.$$

The Fourier transform of $U \in \ell^2$ yields

$$\hat{U}(\xi) = \frac{1}{(2\pi)^{d/2}} \sum_{\nu \in \mathbb{Z}^d} U_\nu e^{i\nu\xi},$$

where

$$\nu\xi = \langle \nu, \xi \rangle = \sum_{k=1}^d \nu_j \xi_j$$

denotes the Euclidean scalar product in $\mathbb{R}^d$.

The stability analysis of system (6.47) with $N \times N$ matrices A_k leads to the following complication. In the univariate case $d = 1$, all coefficient matrices a_j of $C(\lambda, \Delta t)$ are derived from only one matrix A_1, so that in the standard case the matrices a_j are mutually commutable (and thereby simultaneously diagonalisable). For $d > 1$ with non-commutative matrices A_k in (6.47), the matrices a_j are expected to be not simultaneously diagonalisable.

---

[17] In principle, different step widths Δx_j make sense. However, after a transformation $x_j \mapsto \frac{\Delta x_1}{\Delta x_j} x_j$ of the spatial variables we regain a common step size Δx.

6.5.6.2 Time-Dependent Coefficients

The coefficients in the differential equation (6.2) or (6.47) may depend on t:

$$a = a(t) \qquad \text{or} \qquad A_k = A_k(t).$$

In this case the concept of the semigroup of solution operators (cf. §6.2) need to be modified a bit. Instead of $T(t)$, the operator $T(t_1, t_0)$ with $0 \le t_0 \le t_1 \le T$ appears, which describes the transition of an initial value at time t_0 to the solution at time t_1. The semigroup property reads $T(t_2, t_1)T(t_1, t_0) = T(t_2, t_0)$ for $t_0 \le t_1 \le t_2$ and $T(t, t) = I$.

The discretisation yields time-dependent difference schemes $C(t; \lambda, \Delta t)$ at $t = \mu \Delta t$ with $U^\mu = C(\mu \Delta t; \lambda, \Delta t) U^{\mu-1}$:

$$(C(t; \lambda, \Delta t)U)_\nu = \sum_{j \in \mathbb{Z}^d} a_j(t) U_{\nu+j} \qquad \text{for } U \in \ell^p \text{ and } \nu \in \mathbb{Z}^d.$$

In the stability definition (6.23), $C(\lambda, \Delta t)^\mu$ is to be replaced by the product

$$C(t_0 + \mu \Delta t; \lambda, \Delta t) \cdot C(t_0 + (\mu - 1)\Delta t; \lambda, \Delta t) \cdot \ldots \cdot C(t_0 + \Delta t; \lambda, \Delta t)$$

with $0 \le t_0 \le t_0 + \mu \Delta t \le T$.

Criteria 6.26, 6.36 and Lemma 6.34 hold also in the time-dependent case.

6.5.6.3 Spatially Dependent Coefficients

The differential equation may contain spatially dependent coefficients as, e.g., in the differential equation (6.44). Correspondingly, all coefficients $a_j = a_j(x)$ of $C(\lambda, \Delta t)$ may depend on x. Criterion 6.57 explicitly admits variable coefficients in the case of positive difference schemes.

There are criteria using again the function

$$G(x, \xi) := \sum_{j \in \mathbb{Z}} a_j(x) \, \mathrm{e}^{-\mathrm{i}j\xi}, \tag{6.48}$$

which formally corresponds to (6.34). However, G is not the Fourier transform $\mathcal{F}^{-1}C(\lambda, \Delta t)\mathcal{F}$!

Having an operator $C(\lambda, \Delta t)$ with variable coefficients, we can introduce the operator

$$C(x_0; \lambda, \Delta t)$$

by replacing all coefficients $a_j(x)$ by the constant coefficients $a_j(x_0)$. We say that $C(x_0; \lambda, \Delta t)$ is the operator C *frozen at* x_0. $G(x_0, \xi)$ is the Fourier transform of $C(x_0; \lambda, \Delta t)$. This leads to the obvious question as to whether stability of the original scheme $C(\lambda, \Delta t)$ corresponds to the stability of the frozen operator $C(x_0; \lambda, \Delta t)$ at all $x_0 \in \mathbb{R}$. The general answer is negative.

Remark 6.59 (Kreiss [9, §7]). *Stability of $C(x_0; \lambda, \Delta t)$ at all $x_0 \in \mathbb{R}$ is, in general, neither sufficient nor necessary for the stability of $C(\lambda, \Delta t)$.*

Typical sufficient criteria require besides Lipschitz continuity of the coefficients $a_j(\cdot)$ that the scheme be dissipative. Here, $C(\lambda, \Delta t)$ is called *dissipative of order $2r$* ($r \in \mathbb{N}$) if there are constants $\delta, \tau > 0$ such that

$$|\lambda_\nu(x, \Delta t, \xi)| \leq 1 - \delta |\xi|^{2r} \quad \text{for all } x \in \mathbb{R}, \ \Delta t \in (0, \tau), \ |\xi| \leq \pi, \quad (6.49)$$

holds for the eigenvalues λ_ν ($\nu = 1, \ldots, N$) of $G(x, \xi)$.

We hint to a connection between condition (6.49) and the Definition 5.23 of the stability of ordinary differential equations, which requires that zeros λ_ν of ψ with $|\lambda_\nu| = 1$ be simple zeros, while otherwise $|\lambda_\nu| < 1$ must hold. In the case of (6.49), the powers $\|G(x, \xi)^n\|$ must be uniformly bounded, while in the second case the companion matrix must satisfy $\|A^n\| \leq$ const (cf. Remark 5.39). However, the difference is that in the second case only finitely many eigenvalues exist, so that $\max |\lambda_\nu| < 1$ holds for all eigenvalues with $|\lambda_\nu| \neq 1$. In the case of $|\lambda_\nu(x, \Delta t, \xi)|$, the eigenvalues λ_ν are continuous functions of ξ and their absolute value may tend to 1. Condition (6.49) describes quantitatively how fast $\lambda_\nu(x, \Delta t, \xi)$ approaches 1.

The stability result of Kreiss [10] takes the following form (see also [21, §5.4]).

Theorem 6.60. *Suppose the matrices a_j of the hyperbolic system*

$$u_t = \sum_{j=1}^{d} a_j(x) u_{x_j}$$

to be Hermitian, uniformly bounded, and uniformly Lipschitz continuous with respect to x. If the difference scheme is dissipative of order $2r$ and accurate of order $2r - 1$ for some $r > 0$, then it is ℓ^2 stable.

6.5.6.4 Initial-Boundary Value Problem

If we replace the real axis in $\Sigma = [0, T] \times \mathbb{R}$ (cf. (6.3)) by an interval or the half-axis $[0, \infty)$, the new computational domain $\Sigma = [0, T] \times [0, \infty)$ has a non-empty spatial boundary $[0, T] \times \{0\}$. Then the parabolic initial-value problem has to be completed by a boundary condition at $x = 0$; e.g.,

$$u(t, 0) = u_B(t)$$

with some given function u_B to determine a unique solution.

In the case of hyperbolic problems the situation is more involved. The model problem $\frac{\partial}{\partial t} u(t) = a \frac{\partial}{\partial x} u(t)$ in (6.2) requires the same kind of boundary condition as above if $a > 0$. Otherwise, the differential equation is uniquely determined without further condition at $x = 0$. In the case of hyperbolic systems, the number of characteristic directions with positive slope determines the number of necessary boundary conditions.

After the discretisation, the grid $G_{\Delta x}$ from (6.9) has to be replaced by the one-sided infinite grid $G_{\Delta x} = \{x = \nu \Delta x : \nu \in \mathbb{N}_0\}$. The space B becomes $\ell^p = \{U = (U_\nu)_{\nu \in \mathbb{N}_0}\}$, where U_0 is fixed by the boundary conditions. For an analysis we refer to Kreiss [11, 12], Richtmyer–Morton [21, §6], and Kreiss–Wu [13].

6.5.6.5 Multistep Schemes

So far, we have considered one-step methods; i.e., U^{n+1} is computed from U^n. As we already know from ordinary differential equations, instability may arise from improperly designed multistep methods. As an example of a two-step method we discuss the popular *leap-frog scheme*, first applied to the hyperbolic equation $\frac{\partial}{\partial t} u = a \frac{\partial}{\partial x} u$. Both derivatives are discretised by the symmetric difference quotient; i.e.,

$$\frac{\partial}{\partial t} u \approx \frac{1}{2\Delta t}[u(t + \Delta t, x) - u(t - \Delta t, x)] \qquad \text{and}$$

$$\frac{\partial}{\partial x} u \approx \frac{1}{2\Delta x}[u(t, x + \Delta x) - u(t, x - \Delta x)].$$

The resulting two-step method is

$$U_\nu^{\mu+1} = U_\nu^{\mu-1} + \lambda a[U_{\nu+1}^\mu - U_{\nu-1}^\mu]. \tag{6.50}$$

As usual, we need starting values for $\mu = 0$ and $\mu = 1$ to proceed with the leap-frog scheme. The name originates from the fact that the computation of $U_\nu^{\mu+1}$ involves only values

$$U_n^m \quad \text{with } n + m = \nu + \mu + 1 \text{ (modulo 2)}.$$

The grid $\Sigma_{\Delta x}^{\Delta t}$ defined in (6.10) splits into the two colours of the chequer board: $\Sigma_{\text{even}} \dot\cup \Sigma_{\text{odd}}$, where

$$\Sigma_{\text{even}} = \{(n\Delta x, m\Delta t) \in \Sigma_{\Delta x}^{\Delta t} : n + m \text{ even integer}\}.$$

This observation allows us to reduce the computation to Σ_{even}, which halves the cost of the computation.

The stability analysis follows the idea of Remark 5.38b: we formulate the two-step method as a one-step method for $V^\mu := \begin{bmatrix} U^\mu \\ U^{\mu-1} \end{bmatrix}$:

$$\begin{bmatrix} U^{\mu+1} \\ U^\mu \end{bmatrix} = \begin{bmatrix} \lambda a (E_1 - E_{-1}) & 1 \\ 1 & 0 \end{bmatrix} \begin{bmatrix} U^\mu \\ U^{\mu-1} \end{bmatrix};$$

i.e., $V^{\mu+1} = CV^\mu$ with

$$C := \begin{bmatrix} \lambda a \, (E_1 - E_{-1}) & 1 \\ 1 & 0 \end{bmatrix} = \begin{bmatrix} -\lambda a & 0 \\ 0 & 0 \end{bmatrix} E_{-1} + \begin{bmatrix} 0 & 1 \\ 1 & 0 \end{bmatrix} + \begin{bmatrix} \lambda a & 1 \\ 1 & 0 \end{bmatrix} E_1.$$

Here E_j is the shift operator from (6.22). The corresponding characteristic function $G(\xi)$ from (6.34) is now matrix-valued:

$$G(\xi) = \begin{bmatrix} \lambda a \, (\mathrm{e}^{-\mathrm{i}\xi} - \mathrm{e}^{\mathrm{i}\xi}) & 1 \\ 1 & 0 \end{bmatrix} = \begin{bmatrix} -2\mathrm{i}\lambda a \sin \xi & 1 \\ 1 & 0 \end{bmatrix} \quad (0 \le \xi \le 2\pi).$$

One easily checks that $|\lambda a| > 1$ together with $\xi = \pi/2$ (i.e., $\sin \xi = 1$) leads to an eigenvalue $\lambda(\xi)$ with $|\lambda(\xi)| > 1$. Von Neumann's condition in Criterion 6.54a implies that $|\lambda a| \le 1$ is necessary for ℓ^2 stability.

Proposition 6.61. *The leap-frog scheme (6.50) is ℓ^2 stable if and only if $|\lambda a| < 1$.*

Proof. It remains to show that $|\lambda a| < 1$ implies stability. For this purpose we give an explicit description of the powers $G(\xi)^n$. Abbreviate $x := -\lambda a \sin \xi$, so that $G(\xi) = \begin{bmatrix} 2\mathrm{i}x & 1 \\ 1 & 0 \end{bmatrix}$. We claim that

$$G(\xi)^n = \begin{bmatrix} \mathrm{i}^n U_n(x) & \mathrm{i}^{n-1} U_{n-1}(x) \\ \mathrm{i}^{n-1} U_{n-1}(x) & \mathrm{i}^{n-2} U_{n-2}(x) \end{bmatrix} \qquad \text{for } n \ge 1, \qquad (6.51)$$

where U_n are the Chebyshev polynomials of the second kind. These are defined on $[-1,1]$ by

$$U_n(x) := \frac{\sin((n+1)\arccos(x))}{\sqrt{1-x^2}} \qquad (n = -1, 0, 1, \ldots)$$

and satisfy the same three-term recursion

$$U_{n+1}(x) = 2x U_n(x) - U_{n-1}(x)$$

as the Chebyshev polynomials (of the first kind) mentioned in Footnote 7 on page 52. Statement (6.51) holds for $n = 1$. The recursion formula together with $G(\xi)^{n+1} = G(\xi)^n G(\xi)$ proves the induction step.

If $|\lambda a| \le A < 1$, the inequalities $|x| \le A < 1$ and $|\varphi| \le \arccos(A) < 1$ follow, where $\varphi := \arccos(x)$. One verifies that

$$U_n(x) = U_n(\cos \varphi) = \sin((n+1)\varphi)/\sin(\varphi)$$

is bounded by $U_n(A) \approx 1/[2(1 - A)]$ independently of n. Hence $\|G(\xi)^n\|$ is uniformly bounded in n, and ℓ^2 stability follows.

If $A = 1$, the polynomials are bounded by

$$|U_n(x)| \le U_n(0) = n + 1,$$

implying (a rather weak form[18] of) instability. $\square$

---

[18] If the amplification factor is some fixed $\zeta > 1$, the instability effects of $\|C(\lambda, \Delta t)^n\| > \zeta^n$ are easily observed and may lead to an overflow in the very end. On the other hand, we conclude

The original leap-frog scheme of Du Fort–Frankel [4] is applied to the parabolic problem (6.4). Here we start from

$$\frac{\partial}{\partial t} u \approx \frac{1}{2\Delta t}[u(t + \Delta t, x) - u(t - \Delta t, x)] \qquad \text{and}$$

$$\frac{\partial^2}{\partial x^2} u \approx \frac{1}{\Delta x^2}[u(t, x - \Delta x) - 2u(t, x) + u(t, x + \Delta x)].$$

In order to obtain the leap-frog pattern, we replace $2u(t, x)$ by the average $u(t + \Delta t, x) + u(t - \Delta t, x)$. Together with $\lambda = \Delta t/\Delta x^2$ (cf. (6.11)), we obtain the *Du Fort–Frankel scheme*

$$(1 + 2\lambda)U_\nu^{\mu+1} = (1 - 2\lambda)U_\nu^{\mu-1} + 2\lambda[U_{\nu+1}^\mu + U_{\nu-1}^\mu]. \qquad (6.52)$$

The stability analysis can again be based on the representation of $G(\xi)^n$. For $\lambda \le 1/2$, a simpler approach makes use of the fact that the coefficients $\frac{1-2\lambda}{1+2\lambda}$ and $\frac{2\lambda}{1+2\lambda}$ of $U_\nu^{\mu+1}, U_{\nu+1}^\mu, U_{\nu-1}^\mu$ are positive. As in Criterion 6.30, one concludes that the scheme (6.52) is ℓ^∞ stable, which implies ℓ^2 stability.

For $\lambda > 1/2$ we return to the characteristic function

$$G(\xi) = \begin{bmatrix} \frac{2\lambda}{1+2\lambda}\left(e^{-i\xi} + e^{i\xi}\right) & \frac{1-2\lambda}{1+2\lambda} \\ 1 & 0 \end{bmatrix} = \begin{bmatrix} \frac{4\lambda}{1+2\lambda}\cos\xi & \frac{1-2\lambda}{1+2\lambda} \\ 1 & 0 \end{bmatrix} \quad (0 \le \xi \le 2\pi).$$

All eigenvalues are of absolute value ≤ 1. The eigenvalue 1 appears for $\xi = 0$ and is a single one (the other is $\frac{1-2\lambda}{1+2\lambda}$). A double eigenvalue $\sqrt{\frac{2\lambda-1}{2\lambda+1}} < 1$ occurs for $\sin^2\xi = 1/(4\lambda^2)$. Hence, by Lemma 5.28, the powers of $G(\xi)$ are uniformly bounded. This proves the following statement.

Proposition 6.62. *For any fixed* $\lambda = \Delta t/\Delta x^2$, *the Du Fort–Frankel scheme (6.52) is* ℓ^2 *stable.*

As pointed out in [16, §19], it is important to perform the limit $\Delta t, \Delta x \to 0$ in such a way that $\lambda = \Delta t/\Delta x^2$ stays constant, although any value of λ is allowed. If one forms the limit $\Delta t, \Delta x \to 0$ so that the ratio $\mu = \Delta t/\Delta x$ is constant, one obtains the solution of the differential equation

$$u_t = u_{xx} - \mu u_{tt};$$

i.e., consistency is violated.

---

from $\|C(\lambda, \Delta t)^n\| \approx n$ that the result at some fixed $T = n\Delta t > 0$ contains errors, amplified by $T/\Delta t$. If the initial values are such that the consistency error is of the second order $O(\Delta t^2)$, we have still a discretisation error $O(\Delta t)$ at $t = T$. As in §4.6, we have to take into consideration floating point errors, which are also amplified by $\frac{T}{\Delta t}$. Together, we obtain an error $O(\Delta t^2 + \frac{\text{eps}}{\Delta t})$, eps: machine precision, so that we cannot obtain better results than $O(\text{eps}^{2/3})$.

6.5.7 Dissipativity for Parabolic Discretisations

Finally, we discuss the dissipativity (6.49) for discretisations of the heat equation (6.4). Since the solution operator strongly damps high-frequency components, a smoothing effect occurs: initial values u_0, which are only supposed to be continuous (or to belong to L^∞), lead to solutions $u(t)$ of (6.5) which are infinitely often differentiable for all $t > 0$. A corresponding condition for the discrete schemes is condition (6.49), which in this case takes the form

$$|G(\xi)| \leq 1 - \delta\,|\xi|^{2r} \qquad \text{for all } |\xi| \leq \pi, \tag{6.53}$$

since G does not depend on x, and 1×1 matrices coincide with the eigenvalue.

Exercise 6.63. *(a) The simplest scheme (6.18) satisfies (6.53) with the parameters*

$$r = 1,$$
$$\delta = \min\{4\lambda, 2\,(1 - 2\lambda)\}/\pi^2 \quad \text{for } 0 < \lambda \leq 1/2.$$

Dissipativity holds for $0 < \lambda < 1/2$ because of $\delta > 0$, but it fails for $\lambda = 1/2$.

(b) The Crank–Nicolson scheme (which is (6.41) with $\Theta = 1/2$) is dissipative for all $\lambda > 0$ with $r = 1$. For $\lambda \to \infty$, the ℓ^2 stability is uniform (i.e., the stability constant remains bounded for $\lambda \to \infty$), but dissipativity vanishes (i.e., $\delta \to 0$).

6.6 Consistency Versus Stability

Scheme (6.20) is called consistent of order p if p is maximal with the property

$$u(t + \Delta t) - C(\lambda, \Delta t)u(t) = O(\Delta t^{p+1})$$

for all sufficiently smooth solutions u of the differential equations.

The Taylor expansion applied to Example (6.17a) yields

$$u(t + \Delta t, x) - (1 - a\lambda)\,u(t, x) - a\lambda u(t, x + \Delta x) =_{\lambda = \Delta t/\Delta x}$$
$$= u + \Delta t\, u_t + \frac{\Delta t^2}{2} u_{tt} - \left(1 - a\frac{\Delta t}{\Delta x}\right) u - a\frac{\Delta t}{\Delta x}\left[u + \Delta x u_x + \frac{\Delta x^2}{2} u_{xx}\right] + \dots$$
$$= \Delta t\,[u_t - a u_x] + \frac{\Delta t^2}{2}\left[u_{tt} - \frac{\Delta x}{\Delta t} a u_{xx}\right] + \dots$$

Here the functions are to be evaluated at (t, x) and '$\dots$' are higher-order terms. The first bracket vanishes, since u is a solution of $u_t = a u_x$. The second bracket is, in general, different from zero, so that the discretisation (6.17a) has consistency order $p = 1$.

Remark 6.64. *There is an exceptional case in the derivation from above. Applying the differential equation twice, we get*

$$u_{tt} = a(u_x)_t = a(u_t)_x = a(au_x)_x = a^2 u_{xx}.$$

Therefore, $u_{tt} - \frac{\Delta x}{\Delta t} au_{xx} = 0$ proves a higher order of consistency if $\lambda = 1/a$. In fact, the difference scheme becomes $U_\nu^{\mu+1} = U_{\nu+1}^\mu$, and one verifies that the discrete solution coincides with the exact one restricted to the grid (the consistency order may be described by $p = \infty$). This case corresponds to condition (a) in Theorem 6.47.

Also the other schemes (6.17b,c,d) are of first order consistency. A difference scheme of the second order is the following *Lax–Wendroff scheme* (cf. [17, p. 221]):

$$U_\nu^{\mu+1} = \left[\frac{(a\lambda)^2}{2} + \frac{a\lambda}{2}\right] U_{\nu-1}^\mu + \left[1 - (a\lambda)^2\right] U_\nu^\mu + \left[\frac{(a\lambda)^2}{2} - \frac{a\lambda}{2}\right] U_{\nu+1}^\mu. \quad (6.54)$$

Second order consistency is seen from the Taylor expansion

$$u(t + \Delta t, x)$$
$$= u + \Delta t\, u_t + \frac{\Delta t^2}{2} u_{tt} + O(\Delta t^3) = u + a\lambda \Delta x\, u_x + \frac{(a\lambda \Delta x)^2}{2} u_{xx} + O(\Delta t^3)$$
$$= u(t, x) + a\lambda[u(t, x + \Delta x) - u(t, x - \Delta x)]$$
$$\quad + \frac{(a\lambda)^2}{2}[u(t, x + \Delta x) - 2u(t, x) + u(t, x - \Delta x)] + O(\Delta t^3).$$

Lemma 6.65. *The Lax–Wendroff scheme (6.54) is ℓ^2 stable if and only if $|a\lambda| \le 1$.*

Proof. Verify that $G(\xi) = 1 - ia\lambda \sin\xi - (a\lambda)^2 (1 - \cos\xi)$ is the characteristic function. For

$$|G(\xi)|^2 = [1 - \tau(1 - \cos\xi)]^2 + \tau \sin^2\xi \qquad \text{with } \tau := (a\lambda)^2$$

we use $\sin^2\xi = 1 - \cos^2\xi$ and substitute $x := \cos\xi$. Then we have to prove that the polynomial

$$p(x) = [1 - \tau(1 - x)]^2 + \tau(1 - x^2) = 1 + \tau^2 - \tau + 2(\tau - \tau^2)x + (\tau^2 - \tau)x^2$$

remains bounded by 1 for all values $x = \cos\xi \in [-1, 1]$. Inserting $x = -1$ yields $(2\tau - 1)^2$ and proves that $\tau \le 1$ is necessary ($\tau \ge 0$ holds by definition).

Since by definition $|G(\xi)|^2 = p(x) \ge 0$, the maximum is given by

$$\max\{p(1), p(-1)\} = \max\{1, (2\tau - 1)^2\} = 1,$$

where for the last step we used $|a\lambda| \le 1$. Now ℓ^2 stability follows from Theorem 6.44. □

As in §5.5.6, we may ask whether there is a possible conflict between consistency and stability. While in Theorem 5.47 the consistency order p is limited, it is now the parity of p which need to be restricted in certain cases.

We consider the hyperbolic problem $\frac{\partial}{\partial t}u = a\frac{\partial}{\partial x}u$ and a general explicit difference scheme (6.16) with coefficients a_j. Again, we introduce $G(\xi) := \sum_j a_j e^{-ij\xi}$ (cf. (6.34)). Furthermore, we choose the Banach space ℓ^∞ and ask for ℓ^∞ stable schemes. Using Theorem 6.47, Thomée [24] proves the following implication for the consistency order.

Theorem 6.66. *Under the assumption from above, ℓ^∞ stability implies that the consistency order is odd.[19] Furthermore, there are ℓ^∞ stable schemes for any odd order.*

Since the Lax–Wendroff scheme has even consistency order ($p = 2$), it cannot be ℓ^∞ stable. However, the instability is rather weak. Thomée [24] also proves the following two-sided inequality for the Lax–Wendroff scheme:

$$C'n^{1/12} \le \|C(\lambda, \Delta t)^n\|_{\ell^\infty \leftarrow \ell^\infty} \le C''n^{1/6}.$$

References

1. Courant, R., Friedrichs, K.O., Lewy, H.: Über die partiellen Differentialgleichungen der mathematischen Physik. Math. Ann. **100**, 32–74 (1928)
2. Crank, J., Nicolson, P.: A practical method for numerical evaluation of partial differential equations of the heat-conduction type. Proc. Cambridge Phil. Soc. **43**, 50–67 (1947). (reprint in Adv. Comput. Math. **6**, 207–226, 1996)
3. de Moura, C.A., Kubrusly, C.S. (eds.): The Courant–Friedrichs–Lewy (CFL) condition. Springer, New York (2013)
4. Du Fort, E.C., Frankel, S.P.: Stability conditions in the numerical treatment of parabolic differential equations. Math. Tables and Other Aids to Computation **7**, 135–152 (1953)
5. Friedrichs, K.O.: Symmetric hyperbolic linear differential equations. Comm. Pure Appl. Math. **7**, 345–392 (1954)
6. Hackbusch, W.: Iterative Lösung großer schwachbesetzter Gleichungssysteme, 2nd ed. Teubner, Stuttgart (1993)
7. Hackbusch, W.: Iterative solution of large sparse systems of equations. Springer, New York (1994)
8. Hackbusch, W.: Elliptic differential equations. Theory and numerical treatment, *Springer Series in Computational Mathematics*, vol. 18, 2nd ed. Springer, Berlin (2003)
9. Kreiss, H.O.: Über die Stabilitätsdefinition für Differenzengleichungen die partielle Differentialgleichungen approximieren. BIT **2**, 153–181 (1962)
10. Kreiss, H.O.: On difference approximations of the dissipative type for hyperbolic differential equations. Comm. Pure Appl. Math. **17**, 335–353 (1964)
11. Kreiss, H.O.: Difference approximations for the initial-boundary value problem for hyperbolic differential equations. In: D. Greenspan (ed.) Numerical solution of nonlinear differential equations, pp. 141–166. Wiley, New York (1966)
12. Kreiss, H.O.: Initial boundary value problem for hyperbolic systems. Comm. Pure Appl. Math. **23**, 277–298 (1970)
13. Kreiss, H.O., Wu, L.: On the stability definition of difference approximations for the initial boundary value problem. Appl. Numer. Math. **12**, 213–227 (1993)
14. Kröner, D.: Numerical Schemes for Conservation Laws. J. Wiley und Teubner, Stuttgart (1997)

---

[19] This statement does not extend to the parabolic case.

15. Lax, P.D.: Difference approximations of linear differential equations – an operator theoretical approach. In: N. Aronszajn, C.B. Morrey Jr (eds.) Lecture series of the symposium on partial differential equations, pp. 33–66. Dept. Math., Univ. of Kansas (1957)

16. Lax, P.D., Richtmyer, R.D.: Survey of the stability of linear finite difference equations. Comm. Pure Appl. Math. **9**, 267–293 (1965)

17. Lax, P.D., Wendroff, B.: Systems of conservation laws. Comm. Pure Appl. Math. **13**, 217–237 (1960)

18. Lax, P.D., Wendroff, B.: Difference schemes for hyperbolic equations with high order of accuracy. Comm. Pure Appl. Math. **17**, 381–398 (1964)

19. LeVeque, R.J.: Finite Volume Methods for Hyperbolic Problems. Cambridge University Press, Cambridge (2002)

20. Mizohata, S.: The theory of partial differential equations. University Press, Cambridge (1973)

21. Richtmyer, R.D., Morton, K.W.: Difference Methods for Initial-value Problems, 2nd ed. John Wiley & Sons, New York (1967). Reprint by Krieger Publ. Comp., Malabar, Florida, 1994

22. Riesz, F., Sz.-Nagy, B.: Functional Analysis. Dover Publ. Inc, New York (1990)

23. Riesz, M.: Sur les maxima des formes bilinéaires et sur les fonctionelles linéaires. Acta Math. **49**, 465–497 (1926)

24. Thomée, V.: Stability of difference schemes in the maximum-norm. J. Differential Equations **1**, 273–292 (1965)

25. Thorin, O.: An extension of the convexity theorem of M. Riesz. Kungl. Fysiogr. Sällsk. i Lund Förh. **8**(14) (1939)

26. Tomoeda, K.: Stability of Friedrichs's scheme in the maximum norm for hyperbolic systems in one space dimension. Appl. Math. Comput. **7**, 313–320 (1980)

27. Zeidler, E. (ed.): Oxford Users' Guide to Mathematics. Oxford University Press, Oxford (2004)

Chapter 7
Stability for Discretisations of Elliptic Problems

In the previous chapter we treated partial differential equations of hyperbolic and parabolic type. The third type of elliptic differential equations is considered in this chapter. In the hyperbolic and parabolic cases, the solution operator is a mapping from a Banach space B into itself. In the elliptic case the solution operator is given by the inverse differential operator L^{-1}, which is a mapping between *different* spaces. This requires introducing a pair (X, Y) of spaces.

7.1 Elliptic Differential Equations

The prototype of elliptic differential equations is the *Poisson equation*

$$Lu := u_{xx} + u_{yy} = f \quad \text{in } \Omega, \tag{7.1}$$

where Ω is a bounded domain in $\mathbb{R}^2$. To obtain a unique solution, we have to add a *boundary condition* on $\Gamma := \partial\Omega$, e.g., the homogeneous Dirichlet data

$$u = 0 \quad \text{on } \Gamma.$$

Another choice would be the inhomogeneous condition $u = g$ or conditions on (normal) derivatives of u on Γ. The combination of a differential equation (e.g., (7.1)) with a boundary condition is called a *boundary value problem*.

A more general linear differential equations in d variables $x = (x_1, \ldots, x_d)$ is

$$Lu = f \quad \text{with} \quad L = \sum_{i,j=1}^{d} \frac{\partial}{\partial x_i} a_{ij}(x) \frac{\partial}{\partial x_j} + L_0, \tag{7.2}$$

where L_0 may contain further derivatives of order ≤ 1. L is called *uniformly elliptic*, if

W. Hackbusch, *The Concept of Stability in Numerical Mathematics*,
Springer Series in Computational Mathematics 45, DOI 10.1007/978-3-642-39386-0_7,
© Springer-Verlag Berlin Heidelberg 2014

$$\sum_{i,j=1}^{d} a_{ij}(x)\xi_i\xi_j \geq \delta \|\xi\|_2^2 \qquad \text{for all } x \in \Omega \text{ and all } \xi = (\xi_1, \ldots, \xi_d) \in \mathbb{R}^d$$

with a positive constant δ.

We call u a *classical solution* of $Lu = f$ and $u|_\Gamma = 0$, if $u \in C^2(\overline{\Omega}) \cap C_0(\overline{\Omega})$, where

$$C_0(\overline{\Omega}) = \{u \in C(\overline{\Omega}) : u = 0 \text{ on } \Gamma\}.$$

As discussed in §7.4.2, in general, we cannot expect the solution to be a classical one, even for a very smooth boundary.

7.2 Discretisation

As in §6.5.6.1, Ω is covered by a grid. In the case of classical difference schemes, this is a Cartesian grid of step size $h = h_n = \frac{1}{n+1}$, while for finite element discretisations one uses an n-dimensional subspace of functions defined piecewise on a more general triangulation. It is not necessary to require a discretisation for all $n \in \mathbb{N}$. Instead, we assume that there is an infinite subset

$$\mathbb{N}' \subset \mathbb{N},$$

so that the discretisation is defined for all $n \in \mathbb{N}'$.

The Poisson equation (7.1) in the square $\Omega = (0,1)^2$ will serve as a model example. The grid is

$$\Omega_n := \{(\nu h, \mu h) \in \Omega : \nu, \mu = 1, \ldots, n\}.$$

Instead of $u(x, y)$ from (7.1), we are looking for approximations of u at the nodal points $(\nu h, \mu h) \in \Omega_n$:

$$u_{\nu,\mu} \approx u(\nu h, \mu h).$$

In each nodal point $(\nu h, \mu h)$, the second x derivative u_{xx} from (7.1) is replaced by the second divided difference $\frac{1}{h^2}[u_{\nu-1,\mu} - 2u_{\nu,\mu} + u_{\nu+1,\mu}]$. Correspondingly, u_{yy} becomes $\frac{1}{h^2}[u_{\nu,\mu-1} - 2u_{\nu,\mu} + u_{\nu,\mu+1}]$. Together, we obtain the so-called *five-point scheme*:

$$\frac{1}{h^2}[-4u_{\nu,\mu} + u_{\nu-1,\mu} + u_{\nu+1,\mu} + u_{\nu,\mu-1} + u_{\nu,\mu+1}] = f_{\nu,\mu} \quad \text{for all } 1 \leq \nu, \mu \leq n, \tag{7.3}$$

where $f_{\nu,\mu} := f(\nu h, \mu h)$ is the evaluation[1] of the right-hand side f of (7.1). If $\nu = 1$, equation (7.3) contains also the value $u_{\nu-1,\mu} = u_{0,\mu}$. Note that the point $(0, \mu h)$ lies on the boundary $\Gamma_n := \{(x, y) \in \Gamma : x/h, y/h \in \mathbb{Z}\}$, and does not

---

[1] A finite element discretisation using piecewise linear functions in a *regular* triangle grid yields the same matrix, only the right-hand side $\mathbf{f}_h$ consists of integral mean-values $f_{\nu,\mu}$ instead of point evaluations.

belong to Ω_n. Because of the boundary condition $u = 0$, all values $u_{\nu',\mu'}$ in (7.3) with $(\nu'h, \mu'h) \in \Gamma_h$ can be replaced by zero. As a result, one obtains a system of n^2 linear equations for the n^2 unknowns $\{u_{\nu,\mu} : (\nu h, \mu h) \in \Omega_n\}$:

$$\mathbf{L}_n \mathbf{u}_n = \mathbf{f}_n, \tag{7.4}$$

where $\mathbf{u}_n = (u_{\nu,\mu})_{(\nu h, \mu h) \in \Omega_n}$ and $\mathbf{f}_n = (f_{\nu,\mu})_{(\nu h, \mu h) \in \Omega_n}$.

Correspondingly, one can discretise more general differential equations as, e.g., (7.2) in even more complicated domains by difference or finite element methods (cf. [8, 10]).

Remark 7.1. *The matrix $\mathbf{L}_n$ from (7.4) and (7.3) has the following properties:*

(a) $\mathbf{L}_n$ is sparse, in particular, it possesses at most five non-zero entries per row.

(b) $\mathbf{L}_n$ is symmetric.

(c) $-\mathbf{L}_n$ has positive diagonal elements $\frac{4}{h^2}$, while all off-diagonal entries are ≤ 0.

(d) the sum of entries in each row of $-\mathbf{L}_n$ is ≥ 0. More precisely: if $2 \leq \nu, \mu \leq n-1$, the sum is 0; at the corner points $(\nu, \mu) \in \{(1,1), (1,n), (n,1), (n,n)\}$ the sum equals $2/h^2$; for the other points with ν or μ in $\{1, n\}$ the sum is $1/h^2$.

(e) For a concrete representation of the matrix $\mathbf{L}_n$, one must order the components of the vectors $\mathbf{u}_n$, $\mathbf{f}_n$. A possible ordering is the lexicographical one: $(1,1), (2,1), \ldots, (n,1),\ (1,2), \ldots, (n,2),\ \ldots,\ (1,n), \ldots, (n,n)$. In this case, $\mathbf{L}_n$ has the block form

$$\mathbf{L}_n = \frac{1}{h^2}\begin{bmatrix} T & I & & & \\ I & T & I & & \\ & \ddots & \ddots & \ddots & \\ & & I & T & I \\ & & & I & T \end{bmatrix} \quad \text{with} \quad T = \begin{bmatrix} -4 & 1 & & & \\ 1 & -4 & 1 & & \\ & \ddots & \ddots & \ddots & \\ & & 1 & -4 & 1 \\ & & & 1 & -4 \end{bmatrix},$$

where all blocks T, I are of size $n \times n$. Empty blocks and matrix entries are zero.

In the case of the matrix $\mathbf{L}_n$ from (7.3) (but not for any discretisation of elliptic differential equations) $\mathbf{L}_n$ is in particular an M-matrix (cf. [7, §6.4], [8, §4.3]).

Definition 7.2 (M-matrix [17]). *A matrix $A \in \mathbb{R}^{n \times n}$ is called an M-matrix if*

(a) $A_{ii} > 0$ for all $1 \leq i \leq n$,

(b) $A_{ij} \leq 0$ for all $1 \leq i \neq j \leq n$,

(c) A regular, and A^{-1} has only non-negative entries: $\left(A^{-1}\right)_{ij} \geq 0$.

Remark 7.1c shows the properties (a) and (b) of $A = -\mathbf{L}_n$. Remark 7.1d, together with the fact that $\mathbf{L}_n$ is irreducible, describes that $A = -\mathbf{L}_n$ is *irreducibly diagonal dominant*. Irreducibly diagonal dominant matrices with the properties (a) and (b) already possess property (c); i.e., $-\mathbf{L}_n$ is an M-matrix (cf. Hackbusch [7, Theorem 6.4.10b]).

7.3 General Concept

7.3.1 Consistency

$$\begin{array}{ccc} X & \xrightarrow{R_X^n} & X_n \\[4pt] {\scriptstyle\downarrow} L & & {\scriptstyle\downarrow} \mathbf{L}_n \\[4pt] Y & \overset{P_Y^n}{\underset{R_Y^n}{\rightleftarrows}} & Y_n \end{array}$$

The consistency condition must ensure that $\mathbf{L}_n$ is a discretisation of the differential operator L. For this purpose we introduce Banach spaces for the domains and ranges of L and $\mathbf{L}_n$: X, Y, X_n, Y_n, so that

$$L : X \to Y, \qquad \mathbf{L}_n : X_n \to Y_n \qquad (n \in \mathbb{N}') \qquad (7.5\text{a})$$

are continuous mappings. Furthermore, let

$$R_X^n : X \to X_n, \qquad R_Y^n : Y \to Y_n, \qquad \mathbf{f}_n = R_Y^n f \qquad (n \in \mathbb{N}') \qquad (7.5\text{b})$$

be 'restrictions' from the function spaces X, Y into the spaces X_n, Y_n of 'grid functions'. Note that R_Y^n defines the right-hand side $\mathbf{f}_n$ in (7.4). A mapping in the opposite direction is the 'prolongation'

$$P_Y^n : Y_n \to Y \qquad (n \in \mathbb{N}'). \qquad (7.5\text{c})$$

The consistency condition has to relate L and $\mathbf{L}_n$. Since both mappings act in different spaces, the auxiliary mappings R_X^n, R_Y^n, P_Y^n are needed. They permit us to formulate consistency in two versions. Condition (7.6a) measures the consistency error by $\| \cdot \|_{Y_n}$, while (7.6b) uses $\| \cdot \|_Y$:

$$\lim_{\mathbb{N}' \ni n \to \infty} \| (\mathbf{L}_n R_X^n - R_Y^n L) u \|_{Y_n} = 0 \qquad \text{for all } u \in X, \qquad (7.6\text{a})$$

$$\lim_{\mathbb{N}' \ni n \to \infty} \| (P_Y^n \mathbf{L}_n R_X^n - L) u \|_Y = 0 \qquad \text{for all } u \in X. \qquad (7.6\text{b})$$

The conditions (7.6a,b) are almost equivalent. For this purpose, some of the following technical assumptions are of interest:

$$P_Y^n R_Y^n f \to f \qquad\qquad \text{for all } f \in Y, \qquad (7.7\text{a})$$

$$\| P_Y^n \|_{Y \leftarrow Y_n} \le C_P, \qquad\qquad\qquad\qquad\qquad (7.7\text{b})$$

$$\| f_n \|_{Y_n} \le C_P' \| P_Y^n f_n \|_Y \qquad \text{for all } f_h \in Y_h, \qquad (7.7\text{c})$$

$$R_Y^n P_Y^n = I, \qquad\qquad\qquad\qquad\qquad\qquad (7.7\text{d})$$

$$\| R_Y^n \|_{Y_n \leftarrow Y} \le C_R. \qquad\qquad\qquad\qquad\qquad (7.7\text{e})$$

In the case of (7.7d), P_Y^n is a right-inverse of R_Y^n.

Exercise 7.3. *Suppositions (7.7d,e) imply (7.7c) with* $C_P' = C_R$.

Under the conditions (7.7a-c), the consistency formulations (7.6a,b) are equivalent; more precisely:

Proposition 7.4. *(a) If (7.7a,b), then (7.6a) implies (7.6b).*
(b) If (7.7a,c), then (7.6b) implies (7.6a).

Proof. (i) Condition (7.7a) implies $(P_Y^n R_Y^n - I) Lu \to 0$. Using the splitting

$$(P_Y^n \mathbf{L}_n R_X^n - L) u = P_Y^n (\mathbf{L}_n R_X^n - R_Y^n L) u + (P_Y^n R_Y^n - I) Lu,$$

we observe equivalence in the following statements:

$$(P_Y^n \mathbf{L}_n R_X^n - L) u \to 0 \quad \Leftrightarrow \quad P_Y^n (\mathbf{L}_n R_X^n - R_Y^n L) u \to 0.$$

(ii) (7.6a) and (7.7b) imply $P_Y^n (\mathbf{L}_n R_X^n - R_Y^n L) u \to 0$ and, by (i), (7.6b).
(iii) (7.7a) and (7.6b) imply $P_Y^n(\mathbf{L}_n R_X^n - R_Y^n L) u \to 0$. Now assumption (7.7c) proves (7.6a). □

7.3.2 Convergence

In the sequel, a further norm $\|\cdot\|_{\hat{X}_n}$ on X can be fixed, which may be weaker than $\|\cdot\|_{X_n}$ (or equal), but stronger than $\|\cdot\|_{Y_n}$ (or equal):[2]

$$\|\cdot\|_{X_n} \gtrsim \|\cdot\|_{\hat{X}_n} \gtrsim \|\cdot\|_{Y_n}.$$

The discrete space $\hat{X}_n$ should correspond to a space $\hat{X}$ with $X \subset \hat{X} \subset \{u \in Y : u|_\Gamma = 0\}$. The correspondence is expressed by

$$\|R_X^n\|_{\hat{X}_n \leftarrow \hat{X}} \leq \hat{C}_R \qquad \text{for all } n. \tag{7.8}$$

Convergence should express that any solution $u \in X$ of the differential equation $Lu = f$ (with boundary condition $u = 0$) leads to a sequence $\{u_n\}$ of discrete solutions tending to u. Since the discrete grid function $\mathbf{u}_n \in X_n$ and the function $u \in X$ belong to difference spaces, we compare the restriction $R_X^n u \in X_n$ with $\mathbf{u}_n$ and require $\|R_X^n u - \mathbf{u}_n\|_{\hat{X}_n} \to 0$. The corresponding discrete solution takes the form $\mathbf{u}_n = \mathbf{L}_n^{-1} \mathbf{f}_n = \mathbf{L}_n^{-1} R_Y^n f = \mathbf{L}_n^{-1} R_Y^n Lu$. We allow that $\mathbf{L}_n$ is not invertible for finitely many n. Restricting $\mathbb{N}'$ to the remaining set, we suppose in what follows that $\mathbf{L}_n^{-1}$ exists for all $n \in \mathbb{N}'$.

The discussion from above leads to the following definition.

Definition 7.5 (convergence). *With X and $\hat{X}_n$ as above, the sequence $\{\mathbf{L}_n, R_Y^n\}$ defining the discretisation is called* convergent *if*

$$\|(R_X^n - \mathbf{L}_n^{-1} R_Y^n L)u\|_{\hat{X}_n} \to 0 \qquad \text{for all } u \in X \text{ and } \mathbb{N}' \ni n \to \infty. \tag{7.9}$$

An alternative definition of convergence is the following. We replace $u \in X$ by $f = Lu \in \text{range}(L) \subset Y$, and require convergence for all $f \in Y$, i.e., for all solutions $u = L^{-1} f$ with $f \in Y$:

$$\|(R_X^n L^{-1} - \mathbf{L}_n^{-1} R_Y^n)f\|_{\hat{X}_n} \to 0 \qquad \text{for all } f \in Y \text{ and } \mathbb{N}' \ni n \to \infty. \tag{7.10}$$

---

[2] $A_n \gtrsim B_n$ means that there is a constant C, independent of n, such that $A_n \geq C B_n$.

Lemma 7.6. *(a) Statement (7.10) implies (7.9).*
(b) If $L: X \to Y$ is bijective, both characterisations (7.9) and (7.10) are equivalent.

Proof. Let $u \in X$ be arbitrary. For $f := Lu$, $(R_X^n L^{-1} - \mathbf{L}_n^{-1} R_Y^n) f$ from (7.10) equals $(R_X^n - \mathbf{L}_n^{-1} R_Y^n L) u$, proving (7.9), and therefore, Part (a).

In the bijective case, any $f \in Y$ has the representation $f = Lu$ for some $u \in X$. The insertion proves Part (b). $\square$

7.3.3 Stability

The following definition of stability depends on the choice of $\hat{X}_n$.

Definition 7.7. Stability *is characterised by the uniform boundedness of* $\mathbf{L}_n^{-1}$:

$$\sup_{N' \ni n \to \infty} \| \mathbf{L}_n^{-1} \|_{\hat{X}_n \leftarrow Y_n} \leq C_{\text{stab}}. \tag{7.11}$$

Theorem 7.8 (convergence theorem). *Consistency (7.6a) of $u \in X$ and stability (7.11) imply convergence (7.9) to $u \in X$.*

Proof. $(R_X^n - \mathbf{L}_n^{-1} R_Y^n L) u = \mathbf{L}_n^{-1} (\mathbf{L}_n R_X^n - R_Y^n L) u$ yields

$$\| (R_X^n - \mathbf{L}_n^{-1} R_Y^n L) u \|_{\hat{X}_n} \leq \| \mathbf{L}_n^{-1} \|_{X_n \leftarrow Y_n} \| (\mathbf{L}_n R_X^n - R_Y^n L) u \|_{Y_n}$$
$$\leq C_{\text{stab}} \| (\mathbf{L}_n R_X^n - R_Y^n L) u \|_{Y_n} \to 0,$$

proving convergence (7.9). $\square$

Although the setting is similar to the previous chapters, a stability theorem stating that 'convergence implies stability' is, in general, *not* valid, as we shall prove in §7.4.2. However, the stability theorem can be based on the convergence definition in (7.10). Again, the following technical requirement comes into play:

$$\sup_{f \in Y: f_n = R_Y^n f} \frac{\| \mathbf{f}_n \|_{Y_n}}{\| f \|_Y} \geq 1/C_R' \qquad \text{for all } 0 \neq \mathbf{f}_n \in Y_n, n \in \mathbb{N}'. \tag{7.12}$$

Lemma 7.9. *The conditions (7.7b,d) imply (7.12) with $C_R' = C_P$.*

Proof. $f := P_Y^n \mathbf{f}_n$ yields $\sup\{\ldots\} \geq \frac{\| \mathbf{f}_n \|_{Y_n}}{\| P_Y^n \mathbf{f}_n \|_Y} \geq \frac{1}{C_P} \frac{\| \mathbf{f}_n \|_{Y_n}}{\| \mathbf{f}_n \|_{Y_n}} = 1/C_P$. $\square$

Theorem 7.10 (stability theorem). *Suppose (7.8), $\| L^{-1} \|_{\hat{X} \leftarrow Y} \leq C_{L^{-1}}$, and (7.12). Then convergence in the sense of (7.10) implies stability (7.11).*

Proof. Inequality (7.10) describes the point-wise convergence of the operator $R_X^n L^{-1} - \mathbf{L}_n^{-1} R_Y^n$; hence, by Corollary 3.39, the operator norm is uniformly bounded: $\| R_X^n L^{-1} - \mathbf{L}_n^{-1} R_Y^n \|_{\hat{X}_n \leftarrow Y} \leq C$. We conclude that

$$\|\mathbf{L}_n^{-1} R_Y^n\|_{\hat{X}_n \leftarrow Y} \le C + \|R_X^n L^{-1}\|_{\hat{X}_n \leftarrow Y_n} \le C + \hat{C}_R C_{L^{-1}}.$$

Inequality (7.12) implies $\|R_Y^n f\|_{Y_n} \le C_R' \|f\|_Y$. The definition of the operator norm together with (7.12) yields

$$\|\mathbf{L}_n^{-1} R_Y^n\|_{\hat{X}_n \leftarrow Y} = \sup_{0 \neq f \in Y} \frac{\|\mathbf{L}_n^{-1} R_Y^n f\|_{\hat{X}_n}}{\|f\|_Y} = \sup_{0 \neq f \in Y} \frac{\|\mathbf{L}_n^{-1} R_Y^n f\|_{\hat{X}_n}}{\|R_Y^n f\|_{Y_n}} \frac{\|R_Y^n f\|_{Y_n}}{\|f\|_Y}$$

$$= \sup_{0 \neq \mathbf{f}_n \in Y_n} \sup_{f \in Y : \mathbf{f}_n = R_Y^n f} \frac{\|\mathbf{L}_n^{-1} \mathbf{f}_n\|_{\hat{X}_n}}{\|\mathbf{f}_n\|_{Y_n}} \frac{\|\mathbf{f}_n\|_{Y_n}}{\|f\|_Y} \ge \frac{\|\mathbf{L}_n^{-1}\|_{\hat{X}_n \leftarrow Y_n}}{C_R'},$$

proving that $\|\mathbf{L}_n^{-1}\|_{\hat{X}_n \leftarrow Y_n}$ is bounded independently of n. $\quad\square$

Since the convergence definition (7.10) allows us to derive stability, it seems to be the more suitable definition. However, without bijectivity of $L : X \to Y$, definition (7.10) does not lead to a convergence theorem as in Theorem 7.8. In §7.4.2 we shall study bijectivity of L in more detail and gives examples, where bijectivity fails because of range$(L) \subsetneqq Y$. On the other hand, fixing Y and replacing X by $\{L^{-1} f : f \in Y\} \supsetneqq X$ would be rather impractical, since it require a consistency condition involving generalised solutions (cf. §7.4.2).

Only if bijectivity holds, can the equivalence theorem be formulated.

Theorem 7.11 (equivalence theorem). *Suppose that the spaces $\hat{X}$, X, and Y are chosen such that $L : X \to Y$ is bijective, $\|L^{-1}\|_{\hat{X} \leftarrow Y} \le C_{L^{-1}}$, and (7.8) and (7.12) hold. Then convergence (7.9) and stability (7.11) are equivalent.*

Proof. The direction 'stability $\Rightarrow$ convergence' is stated in Theorem 7.8. Because of bijectivity; (7.9) implies (7.10) (cf. Lemma 7.6), and Theorem 7.10 shows '(7.10) $\Rightarrow$ stability'. $\quad\square$

Remark 7.12. *Consistency and convergence can be regarded in two different ways. The previous setting asked for the respective conditions for all $u \in X$. On the other hand, we can consider one particular solution $u = L^{-1} f$. Because of extra smoothness, the consistency error $\varepsilon_n := \|(\mathbf{L}_n R_X^n - R_Y^n L) u\|_{Y_n}$ may behave as $O(n^{-\alpha})$ for some $\alpha > 0$. Then the proof of Theorem 7.8 shows that also the discretisation error $\|(R_X^n - \mathbf{L}_n^{-1} R_Y^n L) u\|_{\hat{X}_n} \le C_{\mathrm{stab}} \varepsilon_n$ is of the same kind.*

7.4 Application to Difference Schemes

7.4.1 Classical Choice of Norms

The concrete choice of the Banach spaces depends on the kind of discretisation. The following spaces correspond to difference methods and *classical* solutions of boundary value problems:

$$
\begin{aligned}
X &= C^2(\overline{\Omega}) \cap C_0(\overline{\Omega}), \\
Y &= C(\overline{\Omega}), \\
X_n &= \mathbb{R}^{n^2}, \quad X_n \ni \mathbf{u}_n = (u_{\nu,\mu})_{(\nu h,\mu h) \in \Omega_n}, \\
Y_n &= \mathbb{R}^{n^2}, \quad Y_n \ni \mathbf{f}_n = (f_{\nu,\mu})_{(\nu h,\mu h) \in \Omega_n},
\end{aligned}
\tag{7.13a}
$$

with the norms[3]

$$
\begin{aligned}
\|u\|_X &= \max\{\|u\|_\infty, \|u_x\|_\infty, \|u_y\|_\infty, \|u_{xx}\|_\infty, \|u_{yy}\|_\infty\}, \\
\|u\|_Y &= \|u\|_\infty := \max\{|u(x,y)| : (x,y) \in \overline{\Omega}\}, \\
\|\mathbf{u}_n\|_{X_n} &= \max\{\|\mathbf{u}_n\|_\infty, \|\partial_x \mathbf{u}_n\|_\infty, \|\partial_y \mathbf{u}_n\|_\infty, \|\partial_{xx} \mathbf{u}_n\|_\infty, \|\partial_{yy} \mathbf{u}_n\|_\infty\}, \\
\|\mathbf{f}_n\|_{Y_n} &= \|\mathbf{f}_n\|_\infty := \max\{|u_{\nu,\mu}| : (\nu h, \mu h) \in \Omega_n\},
\end{aligned}
\tag{7.13b}
$$

where $\partial_x, \partial_y, \partial_{xx}, \partial_{yy}$ are the first and second divided difference quotients with step size $h := 1/n$. The associated restrictions are the point-wise restrictions:

$$
(R_X^n u)_{\nu,\mu} := u(\nu h, \mu h), \qquad (R_Y^n f)_{\nu,\mu} := f(\nu h, \mu h).
$$

The components of $(\mathbf{L}_n R_X^n - R_Y^n L)u$ are

$$
[\partial_{xx} + \partial_{yy}] u(x,y) - [u_{xx}(x,y) + u_{yy}(x,y)] \qquad \text{for } (x,y) \in \Omega_n.
$$

Since, for $u \in X_0 \subset C^2(\overline{\Omega}) \cap C_0(\overline{\Omega})$, the second differences $\partial_{xx} u$ converge uniformly to the second derivative u_{xx}, it follows that $\|(\mathbf{L}_n R_X^n - R_Y^n L)u\|_{X_n} \to 0$ for $n \to 0$. Therefore, the consistency condition (7.6a) is verified.

Consistency follows immediately for $X_0 = X = C^2(\overline{\Omega}) \cap C_0(\overline{\Omega})$, since the difference quotients in $\mathbf{L}_n R_X^n u$ tend uniformly to the derivatives in $R_Y^n Lu$.

The additional space $\hat{X} = C_0(\overline{\Omega})$ is equipped with the maximum norm

$$
\|\mathbf{u}_n\|_{\hat{X}_n} = \|\mathbf{u}_n\|_{Y_n} = \|\mathbf{u}_n\|_\infty.
$$

Therefore, $\|\mathbf{L}_n^{-1}\|_{\hat{X}_n \leftarrow Y_n}$ becomes the row-sum norm $\|\mathbf{L}_n^{-1}\|_\infty$. For the estimate of $\|\mathbf{L}_n^{-1}\|_\infty$ in the model example (7.3), we use that $-\mathbf{L}_n$ is an M-matrix (cf. Definition 7.2). For M-matrices there is a constructive way to determine the row-sum norm of the inverse. In the following lemma, $\mathbf{1}$ is the vector with all entries of value one.

Lemma 7.13. *Let A be an M-matrix and w a vector such that the inequality $Aw \geq \mathbf{1}$ holds component-wise. Then $\|A^{-1}\|_\infty \leq \|w\|_\infty$ holds.*

Proof. For $u \in \mathbb{R}^n$, the vector $(|u_i|)_{i=1}^n$ is denoted by $|u|$. The following inequalities are to be understood component-wise. We have $|u| \leq \|u\|_\infty \mathbf{1} \leq \|u\|_\infty Aw$. Because of the M-matrix property (c), $A^{-1} \geq 0$ holds, so that

$$
|A^{-1} u| \leq A^{-1} |u| \leq A^{-1} \|u\|_\infty Aw = \|u\|_\infty w
$$

and $\|A^{-1} u\|_\infty / \|u\|_\infty \leq \|w\|_\infty$ can be obtained. Therefore, the desired estimate follows: $\|A^{-1}\|_\infty = \sup_{u \neq 0} \|A^{-1} u\|_\infty / \|u\|_\infty \leq \|w\|_\infty$. $\square$

---

[3] Only the norm of Y_h appears in (7.6a) and (7.11).

In the case of $A = -\mathbf{L}_n$ one has to look for a function $w(x, y)$ with $Lw(x, y) \geq 1$. A possible solution is $w(x, y) = \frac{1}{2}x(1 - x)$ with the maximum norm $\|w\|_\infty = 1/8$. Then the point-wise restriction yields the vector $\mathbf{w}_h = R^n_X w$ on the grid Ω_n. Since second differences and second derivatives are identical in the case of a quadratic function, it follows that $(-\mathbf{L}_n)\mathbf{w}_h \geq \mathbb{1}$ and $\|\mathbf{w}_h\|_\infty \leq 1/8$, proving the stability property

$$\|\mathbf{L}_n^{-1}\|_\infty \leq 1/8 \qquad \text{for all } n \in \mathbb{N} \tag{7.14}$$

with $C_{\text{stab}} = 1/8$.

Using the consistency, which is checked above, and the stability result (7.14), we obtain convergence by Theorem 7.8. As discussed in §7.4.2, the given theory does not fully correspond to the previous setting, because convergence does not imply stability.

Another aspect is the order of consistency. So far, only the convergence order $o(1)$ is shown. In general, one likes to show that $\|R^n_X u - \mathbf{u}_n\|_{X_n} = \mathcal{O}(h^\kappa)$ for some $\kappa > 0$. For this purpose, the solution $u \in X$ must have additional smoothness properties: $u \in Z$ for some $Z \subset X = C^2(\overline{\Omega}) \cap C_0(\overline{\Omega})$. Choosing

$$Z := X \cap C^4(\overline{\Omega}),$$

we conclude that $\|(\mathbf{L}_n R^n_X - R^n_Y L)u\|_{Y_n} = \mathcal{O}(h^2)$ for all $u \in Z$. Correspondingly, convergence follows with the same order:

$$\|R^n_X u - \mathbf{u}_n\|_{\hat{X}_n} = \mathcal{O}(h^2).$$

7.4.2 Bijectivity of L

By choosing the spaces X, Y, the operator L is bounded as a mapping from X into Y. Assuming uniqueness of solutions, injectivity of $L \in \mathcal{L}(X, Y)$ holds. This fact guarantees the existence of $L^{-1} \in \mathcal{L}(\text{range}(L), X)$. If, in addition, L is surjective, the inverse L^{-1} is a continuous mapping from Y onto X.

Theorem 7.14. *The differential operators L from (7.1) or (7.2) belong to $\mathcal{L}(C^2(\overline{\Omega}) \cap C_0(\overline{\Omega}), C(\overline{\Omega}))$, but are not surjective.*

The proof is given in [8, 10, Theorem 3.2.7]. The consequences are:

1. The definitions (7.9) and (7.10) of convergence are not equivalent.
2. If L is injective, L^{-1} exists as bounded mapping from $C(\overline{\Omega})$, e.g., into $\hat{X} = C^1(\overline{\Omega}) \cap C_0(\overline{\Omega})$ or $\hat{X} = C_0(\overline{\Omega})$, but it does not belong to the space $\mathcal{L}(C(\overline{\Omega}), C^2(\overline{\Omega}) \cap C_0(\overline{\Omega}))$.

Concerning the first statement, note that the characterisation (7.9) refers to all solutions in $C^2(\overline{\Omega}) \cap C_0(\overline{\Omega})$, whereas (7.10) refers to all solutions $u = L^{-1}f$ for $f \in C(\overline{\Omega})$.

For the proof of the second statement, we have to introduce the generalised solutions $u = L^{-1}f$, which do not belong to $C^2(\overline{\Omega}) \cap C_0(\overline{\Omega})$. For this purpose, one can use Green's representation

$$u(x) = \int_\Omega G(x,y)f(y)\mathrm{d}y. \tag{7.15}$$

The Green function $G(\cdot,\cdot)$ satisfies $LG(\cdot,y)=0$ for all $y\in\Omega$ and $G(x,\cdot)=0$ for $x\in\Gamma$. In the case of Poisson's equation (7.1) in the circle $\Omega = \{\|x\| < 1\} \subset \mathbb{R}^2$,

$$G(x,y) = -\frac{1}{2\pi}\left[\log\|x-y\| - \log\left(\|x\|\,\Big\|y - \frac{1}{\|x\|^2}x\Big\|\right)\right]$$

holds (cf. §2.2 in [8, 10]). Since G as well as $\frac{\partial}{\partial x_i}G(x,y)$ has an integrable singularity, the integral in (7.15) exists and defines a function in $\hat{X} = C^1(\overline{\Omega}) \cap C_0(\overline{\Omega})$. In particular, $\hat{X} = C_0(\overline{\Omega})$ leads to the (finite) operator norm

$$\left\|L^{-1}\right\|_{C(\overline{\Omega})\leftarrow C(\overline{\Omega})} = \int_\Omega |G(x,y)|\,\mathrm{d}y.$$

If f is Hölder continuous, the function u defined in (7.15) belongs to C^2 and is a solution of $Lu = f$. This proves that (7.15) defines generalised solutions in $\hat{X}$ and that $L^{-1} \in \mathcal{L}(C(\overline{\Omega}),\hat{X})$.

To apply the equivalence Theorem 7.11, we have to look for spaces X, Y, such that $L : X \to Y$ is bijective. An example of classical function spaces are the Hölder spaces $Y = C^\lambda(\overline{\Omega})$, $0 < \lambda < 1$, and $X = C^{2+\lambda}(\overline{\Omega})$. The (unusual) discrete norm on Y_n is defined by the maximum over all $|f(x)|$, $x \in \Omega_n$, and $|f(x) - f(y)| / \|x - y\|^\lambda$, $x, y \in \Omega_n$.

A canonical choice of X, Y with a bijective $L : X \to Y$, obtained from the variational setting, are the Sobolev spaces $X = H_0^1(\Omega)$ and $Y = H^{-1}(\Omega)$ (cf. §7.5). Under suitable conditions (e.g., smooth coefficients a_{ij} in (7.2) and convex Ω), also the choices of $X = H^2(\Omega) \cap H_0^1(\Omega)$ and $Y = L^2(\Omega)$ are valid.

The previous convexity condition excludes domains Ω with reentrant corners. A class of domains allowing reentrant corners are *Lipschitz domains* (i.e., there must be a Lipschitz continuous parametrisation of Γ). Nečas [15] proves the following regularity result.

Theorem 7.15. *Let Ω be a bounded Lipschitz domain, and suppose that L is uniformly elliptic with Hölder continuous coefficients $a_{ij} \in C^t(\overline{\Omega})$ of order $t \in (0, \frac{1}{2}]$. Then $L : X \to Y$ and $L^{-1} : Y \to X$ are bounded for the Sobolev spaces[4]*

$$X = H_0^{1+s}(\Omega), \; Y = H^{-1+s}(\Omega) \qquad \text{for any } s \text{ with } 0 \le s < t \le 1/2.$$

Note that $s = 0$ yields the trivial statement $L : H_0^1(\Omega) \leftrightarrows H^{-1}(\Omega)$. By Theorem 7.15, smooth functions f produce solutions in the space $H_0^{3/2-\varepsilon}(\Omega)$ for all $\varepsilon > 0$.

---

[4] The Sobolev spaces $H^t(\Omega)$ and $H_0^t(\Omega)$ are introduced in [8, 10].

It is possible to discretise the boundary value problem by a difference scheme and to analyse $\mathbf{L}_n$ with respect to discrete Sobolev spaces X_n and Y_n of the respective order $1 + s$ and $-1 + s$ (cf. Hackbusch [6]). Then, the stability estimate $\|\mathbf{L}_n^{-1}\|_{Y_n \leftarrow X_n} \leq C_{\text{stab}}$ holds as well as consistency $\| (\mathbf{L}_n R_X^n - R_Y^n L) u\|_{Y_n} \to 0$ for all $u \in X = H_0^{1+s}(\Omega)$. Regularity properties of difference schemes are also discussed in Jovanovič–Süli [13].

7.5 Finite Element Discretisation

7.5.1 Variational Problem

The variation formulation of the boundary value problem (7.1) and (7.2) is based on a bilinear form:

$$a(u, v) = -\int_\Omega f(x)v(x)\mathrm{d}x \qquad \text{for all } v \in H_0^1(\Omega), \text{ where} \qquad (7.16a)$$

$$a(u, v) := \int_\Omega \sum_{j=1}^d \frac{\partial u(x)}{\partial x_j} \frac{\partial v(x)}{\partial x_j} \mathrm{d}\dot{x}. \qquad (7.16b)$$

The Sobolev space $H_0^1(\Omega)$ can be understood as the completion of $C^1(\Omega) \cap C_0(\Omega)$ with respect to the norm

$$\|u\|_{H^1} := \sqrt{\int_\Omega \left[|u(x)|^2 + \sum_{j=1}^d \left| \frac{\partial}{\partial x_j} u(x) \right|^2 \right] \mathrm{d}x}$$

(cf. §6.2 in [8, 10]). The dual space $H^{-1}(\Omega)$ consists of all functionals with finite norm

$$\|f\|_{H^{-1}} = \sup\{|f(v)| : v \in H_0^1(\Omega), \|v\|_{H^1} = 1\}.$$

The second embedding in $H_0^1(\Omega) \subset L^2(\Omega) \subset H^{-1}(\Omega)$ is based on the identification of functions $f \in L^2(\Omega)$ with the functional $f(v) := \int_\Omega f(x)v(x)\mathrm{d}x$.

If the bilinear form $a(\cdot, \cdot)$ is bounded, i.e.,

$$C_a := \sup\{a(u, v) : u, v \in H_0^1(\Omega), \|u\|_{H^1} = \|v\|_{H^1} = 1\} < \infty, \qquad (7.17)$$

and problem (7.16a) is identical to the abstract equation

$$Au = f \qquad \text{with } A \in \mathcal{L}(H_0^1(\Omega), H^{-1}(\Omega)), \quad u \in H_0^1(\Omega), \quad f \in H^{-1}(\Omega).$$

If, in general, a is a bilinear form on $X \times X$, the operator $A \in \mathcal{L}(X, X^*)$ (X^* dual space of X) is defined by

$$(Au)(v) := a(u, v) \qquad \text{for all } u, v \in X. \qquad (7.18)$$

The norm $\|A\|_{H^1 \leftarrow H^{-1}}$ coincides with C_a from above. The particular example (7.16b) satisfies $C_a < 1$.

If $A^{-1} \in \mathcal{L}(H^{-1}(\Omega), H_0^1(\Omega))$ exists, $u := A^{-1}f$ is the desired solution (it is called a 'weak solution'). The existence of A^{-1} can be expressed by the so-called inf-sup conditions for the bilinear form a (more precisely, the inf-sup expression is an equivalent formulation of $1/\|A^{-1}\|_{H^1 \leftarrow H^{-1}}$; cf. §6.5 in [8, 10]).

A very convenient, *sufficient* condition is the $H_0^1(\Omega)$-coercivity[5]

$$a(u, u) \geq \varepsilon_{\mathrm{co}} \|u\|_{H^1}^2 \quad \text{with } \varepsilon_{\mathrm{co}} > 0 \text{ for all } u \in H_0^1(\Omega). \tag{7.19}$$

Above, we used the spaces X, Y together with the differential operator L. Now L will be replaced by A, involving the spaces

$$X := H_0^1(\Omega), \quad Y := H^{-1}(\Omega).$$

Note that $Y = X^*$ is the dual space with the dual norm

$$\|f\|_{X^*} := \sup\{|f(u)| : u \in X, \|u\|_X \leq 1\}.$$

As (7.19) is valid for (7.16b), $A \in \mathcal{L}(X, Y)$ is bijective; i.e., $A^{-1} \in \mathcal{L}(Y, X)$ exists.

7.5.2 Galerkin Discretisation

The Galerkin discretisation is completely characterised by a subspace $U_n \subset X$ of dimension n. We consider a family $\{U_n\}_{n \in \mathbb{N}'}$, where $\mathbb{N}' \subset \mathbb{N}$ is an infinite subset. The finite element method is a particular example, where U_n consists of piecewise polynomials. The vector space U_n can be equipped with different norms. We set $X_n := U_n$ equipped with the norm $\|\cdot\|_{H^1}$. The discrete solution $u_n \in X_n$ is defined by

$$a(u_n, v_n) = -\int_\Omega f(x)v_n(x)\mathrm{d}x \quad \text{for all } v_n \in X_n.$$

The restriction of a bilinear form $a(\cdot, \cdot)$ on $X \times X$ to a subspace $X_n \times X_n$ yields again a bilinear form $a_n(\cdot, \cdot)$ on $X_n \times X_n$. If a is bounded by C_a (cf. (7.17)), the bound C_{a_n} of a_n satisfies $C_{a_n} \leq C_a$. In the same way as $A \in \mathcal{L}(X, X^*)$ is uniquely determined by a, the bilinear form a_n corresponds to an operator

$$A_n \in \mathcal{L}(X_n, X_n^*). \tag{7.20}$$

Note that X_n^* is U_n equipped with the norm dual to X_n.

---

[5] The algebraic counterpart is the following statement: if $A + A^{\mathsf{H}}$ is positive definite, then A is regular. More precisely, $A + A^{\mathsf{H}} \geq 2\varepsilon_{\mathrm{co}}I$ implies $\|A^{-1}\|_2 \leq 1/\varepsilon_{\mathrm{co}}$. In the case of $A = A^{\mathsf{H}}$, $\|A^{-1}\|_2 = 1/\varepsilon_{\mathrm{co}}$ holds.

Exercise 7.16. *The dual norm of X_n^* is not the restriction of the dual norm of X^* to U_n. Prove that $\|f_n\|_{X_n^*} \leq \|f_n\|_{X^*}$ for $f_n \in X_n^*$.*

Remark 7.17. *Let $X_n \subset X$. If $a(\cdot,\cdot)$ is X-coercive with constant ε (cf. (7.19)), $a_n(\cdot,\cdot)$ is X_n-coercive with a constant $\varepsilon_{n,\mathrm{co}}$ satisfying $\varepsilon_{n,\mathrm{co}} \geq \varepsilon_{\mathrm{co}}$. As a consequence, X-coercive bilinear forms $a(\cdot,\cdot)$ lead to stable Galerkin approximations such that $A_n^{-1} \in \mathcal{L}(X_n^*, X_n)$ exists. In the symmetric case,*

$$\|A_n^{-1}\|_{X_n \leftarrow X_n^*} \leq \|A^{-1}\|_{X \leftarrow X^*}$$

holds.

7.5.3 Consistency

The main *consistency condition* is described in terms of the subspaces U_n: the distance between U_n and any $u \in X$ has to satisfy

$$\inf\{\|u - v_n\|_X : v_n \in U_n\} \to 0 \qquad \text{for } n \to \infty \text{ and all } u \in X.$$

An equivalent statement is the point-wise convergence

$$\|u - \Pi_n u\|_X \to 0 \qquad \text{for all } u \in X \text{ as } n \to \infty, \tag{7.21}$$

where $\Pi_n : X \to X_n$ is the X-orthogonal projection onto X_n. The projection may be considered as an element of $\mathcal{L}(X, X)$ as well as $\mathcal{L}(X, X_n)$. The dual operator Π_n^* then belongs to $\mathcal{L}(X^*, X^*)$ as well as $\mathcal{L}(X_n^*, X^*)$.

Remark 7.18. *(a) The Riesz isomorphism[6] $J : X \to X^*$ is a unitary operator; i.e., $J^* = J^{-1}$. For any $f \in X^*$, the action $f(u)$, $u \in X$, can be described by the scalar product $\langle J^* f, u \rangle_X$. X^* is a Hilbert space with the scalar product*

$$\langle f, g \rangle_{X^*} := \langle J^* f, J^* g \rangle_X .$$

(b) Let $X = X_n \oplus X_n^\perp$ be an X-orthogonal decomposition and set $Y_n := J X_n$, $Y_n^\perp := J X_n^\perp$. Then $X^* = Y_n \oplus Y_n^\perp$ is an X^*-orthogonal decomposition.
(c) The X-orthogonal projection $\Pi_n \in \mathcal{L}(X, X)$ onto X_n satisfies

$$J \Pi_n = \Pi_n^* J;$$

i.e., the dual projection has the representation $\Pi_n^* = J \Pi_n J^* = J \Pi_n J^{-1}$.
(d) Property (7.21) is equivalent to

$$\|v - \Pi_n^* v\|_{X^*} \to 0 \qquad \text{for all } v \in X^* \text{ as } n \to \infty. \tag{7.22}$$

---

[6] Also called the Fréchet–Riesz isomorphism, since both, Fréchet [4] and Frigyes Riesz [19] published their results in 1907. See also [20, II.30].

Proof. For Part (b) write $y_n \in Y_n$ and $y_n^{\perp} \in Y_n^{\perp}$ as $y_n = Jx_n$ and $y_n^{\perp} = Jx_n^{\perp}$ for suitable $x_n \in X_n$ and $x_n^{\perp} \in X_n^{\perp}$. Then

$$\langle y_n, y_n^{\perp} \rangle_{X^*} = \langle J^* y_n, J^* y_n^{\perp} \rangle_X = \langle J^{-1} y_n, J^{-1} y_n^{\perp} \rangle_X = \langle x_n, x_n^{\perp} \rangle_X = 0.$$

For Part (c) verify that Π_n^* maps $y = y_n + y_n^{\perp}$ into y_n (with notations $y_n \in Y_n$ and $y_n^{\perp} \in Y_n^{\perp}$ as in Part (b)).

For Part (d) with $v = Ju$ use $\|v - \Pi_n^* v\|_{X^*} = \|(I - \Pi_n^*) Ju\|_{X^*} =_{\text{Part (c)}}$ $\|J(I - \Pi_n) u\|_{X^*} = \|(I - \Pi_n) u\|_X$. □

We recall the operators A from (7.18) and A_n from (7.20).

Remark 7.19. *The relation between $A \in \mathcal{L}(X, X^*)$ and $A_n \in \mathcal{L}(X_n, X_n^*)$ is given by*

$$A_n = \Pi_n^* A \Pi_n.$$

Proof. This follows from $\langle A_n u_n, v_n \rangle_{X_n^* \times X_n} = a(u_n, v_n) = a(\Pi_n u_n, \Pi_n v_n) = \langle A \Pi_n u_n, \Pi_n v_n \rangle_{X^* \times X} = \langle \Pi_n^* A \Pi_n u_n, v_n \rangle_{X^* \times X}$ for all $u_n, v_n \in X_n$. □

The canonical choice of the space Y_n is X_n^*. The mappings between X, Y, X_n, Y_n are as follows:

$$Y := X^*, \quad Y_n := X_n^*, \tag{7.23a}$$
$$R_X^n = \Pi_n : X \to X_n \quad X\text{-orthogonal projection onto } X_n, \tag{7.23b}$$
$$R_Y^n : Y \to Y_n \text{ restriction to } X_n \subset X; \text{ i.e., } f_n = R_Y^n f = f|_{X_n} \in X_n^*, \tag{7.23c}$$
$$P_Y^n := (R_X^n)^* = \Pi_n^* = J \Pi_n J^* : Y_n \to Y. \tag{7.23d}$$

We consider R_X^n as a mapping onto X_n (not into X). Concerning R_Y^n, note that the mapping $f \in Y = X^*$ can be restricted to the subspace $X_n \subset X$. For $P_Y^n = \Pi_n^*$ compare Remark 7.19a.

Lemma 7.20. *Assumption (7.21) implies the consistency statement (7.6a).*

Proof. Application of the functional $(A_n R_X^n - R_Y^n A) u$ to $v_n \in X_n$ yields

$$[(A_n R_X^n - R_Y^n A) u](v_n) = a_n(R_X^n u, v_n) - a(u, v_n) = a(\Pi_n u, v_n) - a(u, v_n)$$
$$= a(\Pi_n u - u, v_n),$$

proving $\|(\mathbf{L}_n R_X^n - R_Y^n L) u\|_{Y_n} \leq C_a \|u - \Pi_n u\|_X \to 0$ with C_a from (7.17) because of (7.21). □

Next, we verify that all conditions (7.7a–e) are valid. The constants are $C_P = C_P' = C_R = 1$. Concerning the proof of (7.7a), we note that

$$P_Y^n R_Y^n f - f = \Pi_n^* f - f \underset{(7.22)}{\to} 0.$$

Hence, the consistency statement (7.6a) is equivalent to (7.6b) (cf. Proposition 7.4a), which reads $(P_Y^n A_n R_X^n - A) u = (\Pi_n^* A_n \Pi_n - A) u \to 0$ in Y.

7.5.4 Convergence and Stability

We assume that A is invertible; i.e., the variational problem (7.16a) is uniquely solvable. First we set $\hat{X}_n := X_n$ with norm $\| \cdot \|_{\hat{X}_n} = \| \cdot \|_{X_n}$. For further conclusions, we have to ensure stability. For simplicity, we assume that the bilinear form is X-coercive (cf. (7.19)). Then, as stated in Remark 7.17, $\varepsilon_{n,co} \geq \varepsilon_{co}$ proves stability.

From consistency (7.6a), using Theorem 7.8, we infer convergence in both (7.9) and (7.10), since $A : X \to Y$ is bijective (cf. Lemma 7.6b). The convergence statement (7.10) takes the form

$$(R_X^n A^{-1} - A_n^{-1} R_Y^n)f = (\Pi_n A^{-1} - A_n^{-1})f$$
$$= \Pi_n u - u_n \to 0 \qquad \text{in } X \text{ for } \mathbb{N}' \ni n \to \infty,$$

where u and u_n are the exact and discrete solutions, respectively. This is not the standard Galerkin convergence statement

$$u - u_n \to 0 \qquad \text{in } X \text{ for } \mathbb{N}' \ni n \to \infty,$$

but the latter statement follows immediately from condition (7.21): $u - \Pi_n u \to 0$ in X.

7.5.5 Quantitative Discretisation Error and Regularity

Finite element discretisation uses the subspace $U_n \subset X = H_0^1(\Omega)$ of, e.g., piecewise linear functions on a triangulation. Let $h = h_n$ be the largest diameter of the involved triangles (or other elements, e.g., tetrahedra in the three-dimensional case). In the standard case, $h_n = O(n^{-1/d})$ is to be expected (d: spatial dimension, $\Omega \subset \mathbb{R}^d$), provided that n is related to the dimension: $n = \dim(U_n)$.

The consistency error is bounded by $C_a \|u - \Pi_n u\|_X$ as shown by the proof of Lemma 7.20. According to Remark 7.12, the discretisation error $\|u - u_n\|_X$ can be estimated by[7] $(1 + C_{\text{stab}} C_a) \|u - \Pi_n u\|_X$. While $u \in X = H_0^1(\Omega)$ allows only point-wise convergence, quantitative error bound can be expected for smoother functions u. Assume, e.g., $u \in H^2(\Omega) \cap H_0^1(\Omega)$. Then the finite element error can be proved to be $\|u - u_n\|_{H_0^1(\Omega)} = O(h)$. More generally, finite elements of degree p lead to an error[8]

$$\|u - u_n\|_{H_0^1(\Omega)} = O(h_n^{\min(p,t-1)}), \quad \text{if } u \in H^t(\Omega) \cap H_0^1(\Omega), \ t > 1. \quad (7.24)$$

The property $u \in H^t(\Omega)$ may hold by accident or for systematic reasons ('*regularity*' properties'). Solvability of the boundary value problem is equivalent

---

[7] This is the result of Cea's lemma, which is usually proved differently (cf. Theorem 8.2.1 in [8, 10]).

[8] See Footnote 4 on page 148.

to $A^{-1} \in \mathcal{L}(H^{-1}(\Omega), H_0^1(\Omega))$ (cf. §7.5.1). It describes that the solution process increases the degree of smoothness by 2. One may ask whether a similar statement

$$A^{-1} \in \mathcal{L}(H^{t-2}(\Omega), H^t(\Omega) \cap H_0^1(\Omega))$$

holds for certain $t > 1$. In this case, the problem is called t-regular. Such regularity statements depend on the smoothness of the coefficients of the differential operator (7.2) and on the smoothness of the boundary. Theorem 7.15 yields t-regularity for $1 < t < 3/2$ under the condition that Ω is a Lipschitz domain. For convex domains (or domains that are smooth images of convex domains) and sufficiently smooth coefficients, the problem is 2-regular:

$$A^{-1} \in \mathcal{L}(L^2(\Omega), H^2(\Omega) \cap H_0^1(\Omega)).$$

Piecewise linear finite elements are of degree $p = 1$. Assuming 2-regularity, $f \in L^2(\Omega)$ ensures $u \in H^2(\Omega) \cap H_0^1(\Omega)$, so that $\|u - u_n\|_{H_0^1(\Omega)} = C_2 h_n \|f\|_{L^2(\Omega)}$ follows from (7.24).

7.5.6 L^2 Error

The adjoint operator $A^* \in \mathcal{L}(X, X^*)$ belongs to the adjoint bilinear form a^* defined by $a^*(u, v) := a(v, u)$. In this section we assume that A^* is 2-regular.

The representation $u - u_n = A^{-1} f - A_n^{-1} \Pi_n f = (A^{-1} - A_n^{-1} \Pi_n) f$ shows that

$$E_n := A^{-1} - A_n^{-1} \Pi_n = A^{-1} - \Pi_n A_n^{-1} \Pi_n$$

can be considered as *error operator* with the property

$$\|E_n f\|_{H_0^1(\Omega)} \le C_2 h_n \|f\|_{L^2(\Omega)},$$

where C_2 depends on details of the finite element triangulation (cf. [8, 10]).

Lemma 7.21. $E_n A E_n = E_n$ *holds for Galerkin discretisation.*

Proof. The identity

$$E_n A E_n = (A^{-1} - A_n^{-1} \Pi_n) A (A^{-1} - \Pi_n A_n^{-1} \Pi_n)$$
$$= A^{-1} - 2 A_n^{-1} \Pi_n + A_n^{-1} \Pi_n A \Pi_n A_n^{-1} \Pi_n = E_n$$

follows from $\Pi_n A \Pi_n = A_n$. $\square$

Theorem 7.22. *Assume that A^* is 2-regular, and $\|E_n^* g\|_{H_0^1(\Omega)} \le C_2 h_n \|g\|_{L^2(\Omega)}$. Then*

$$\|u - u_n\|_{L^2(\Omega)} = \|E_n f\|_{L^2(\Omega)} \le C_a C_2 h_n \|E_n f\|_{H_0^1(\Omega)}.$$

Proof. For any $g \in L^2(\Omega)$ with $\|g\|_{L^2(\Omega)} = 1$ and $f \in X^*$, we have

$$\langle E_n f, g \rangle_{X \times X^*} \underset{\text{Lemma 7.21}}{=} \langle E_n A E_n f, g \rangle_{X \times X^*} = \langle A E_n f, E_n^* g \rangle_{X^* \times X}$$
$$= a(E_n f, E_n^* g),$$
$$\left| \langle E_n f, g \rangle_{X \times X^*} \right| \leq C_a \|E_n f\|_X \|E_n^* g\|_X \leq C_a C_2 h_n \|E_n f\|_X.$$

By identification $L^2(\Omega) = (L^2(\Omega))^*$, we can continue with

$$\langle Ef, g \rangle_{X \times X^*} = g(Ef) = (Ef, g)_{L^2(\Omega)}$$

for $g \in L^2(\Omega)$. Since

$$\|E_n f\|_{L^2(\Omega)} = \max\{| (Ef, g)_{L^2(\Omega)} | : g \in L^2(\Omega), \|g\|_{L^2(\Omega)} = 1\},$$

the assertion follows. $\square$

Theorem 7.22 states that the L^2 error is by one factor of h_n better than the H^1 error $\|E_n f\|_{H^1_0(\Omega)} = \|u - u_n\|_{H^1_0(\Omega)}$. This result, which traces back to Aubin [1] and Nitsche [16], is usually proved differently, making indirect use of Lemma 7.21.

7.5.7 Stability of Saddle Point Problems

If coercivity (7.19) holds and if a is symmetric: $a(u, v) = a(v, u)$, the variational formulation (7.16a) is equivalent to the minimisation of $J(u) := \frac{1}{2}a(u, u) - f(u)$. Quite another type is the following saddle point problem. We are looking for functions $v \in V$ and $w \in W$ (V, W Hilbert spaces) satisfying

$$\begin{aligned} a(v, v') + b(w, v') &= f_1(v') && \text{for all } v' \in V, \\ b(y, w') &= f_2(w') && \text{for all } w' \in W, \end{aligned} \tag{7.25}$$

with bilinear forms $a : V \times V \to \mathbb{R}$ and $b : W \times V \to \mathbb{R}$.

An example of the general form (7.25) is Stokes' problem, where

$$V = (H^1_0(\Omega))^d, \qquad W = \{f \in L^2(\Omega) : \int_\Omega f \mathrm{d}\xi = 0\},$$

$$a(v, v') = \int_\Omega \langle \nabla v, \nabla x \rangle \, \mathrm{d}\xi, \quad b(w, v') = \int_\Omega w \operatorname{div} v' \mathrm{d}\xi. \tag{7.26}$$

Define

$$J(v, w) := \tfrac{1}{2}a(v, v) + b(w, v) - f_1(v) - f_2(w).$$

The following saddle point properties are proved in [8, 10, Theorem 12.2.4].

Theorem 7.23. *Let* $a(\cdot, \cdot)$ *be symmetric and coercive.* $(v, w) \in V \times W$ *is a solution of (7.25), if and only if*

$$J(v, w') \leq J(v, w) \leq J(v', w) \qquad \text{for all } v' \in V, w' \in W.$$

Furthermore,

$$J(v, w) = \min_{v' \in V} J(v', w) = \max_{w' \in W} \min_{v' \in V} J(v', w').$$

The bilinear forms a and b in (7.25) give rise to operators $A \in \mathcal{L}(V, V^*)$ and $B \in \mathcal{L}(W, V^*)$. The variational setting (7.25) is equivalent to the operator equation

$$\begin{bmatrix} A & B \\ B^* & 0 \end{bmatrix} \begin{bmatrix} v \\ w \end{bmatrix} = \begin{bmatrix} f_1 \\ f_2 \end{bmatrix}.$$

Solvability of the problem is equivalent to the existence of the inverse of

$$C := \begin{bmatrix} A & B \\ B^* & 0 \end{bmatrix}.$$

Since C is symmetric, its eigenvalues are real. In the case of a symmetric and coercive $a(\cdot, \cdot)$, all eigenvalues of the operator A are positive. However, C is indefinite; i.e., it has positive and negative eigenvalues. One can show that the discrete operator A_n has eigenvalues larger or equal to those of A (Lemma 7.17 states this result for the smallest eigenvalue). In the indefinite case, such a simple result is not available. To the contrary, a wrong type of discretisation can easily generate zero eigenvalues of C_n; i.e., the discrete problem is not solvable.

The following counterexample reveals a paradoxical situation. We choose subspaces $V_n \subset V$ and $W_n \subset W$ such that

$$\dim W_n > \dim V_n.$$

Seemingly, a high-dimensional subspace W_n should improve the approximation of the component w, but instead, it spoils the whole system. For a proof consider the block

$$\begin{bmatrix} B_n \\ 0 \end{bmatrix}.$$

In the best case, this part has rank $\dim V_n$ (number of rows of B_n). Since the matrix C_n is of size $N_n \times N_n$ with $N_n := \dim W_n + \dim V_n$, its rank is bounded by $2 \dim V_n < N_n$. Therefore, at least $\dim W_n - \dim V_n > 0$ vanishing eigenvalues must exist.

An equivalent statement for the stability

$$\|C_n^{-1}\|_{(V_n \times W_n) \leftarrow (V_n^* \times W_n^*)} \leq C_{\text{stab}}$$

is the following inf-sup condition, which is also called the *Babuška–Brezzi condition:*

$$\inf_{v_0 \in V_{0,n} \text{ with } \|v_0\|_V = 1} \sup_{x_0 \in V_{0,n} \text{ with } \|x_0\|_V = 1} |a(v_0, x_0)| \geq \alpha > 0, \tag{7.27a}$$

$$\sup_{x_0 \in V_{0,n} \text{ with } \|x_0\|_V = 1} |a(x_0, v_0)| > 0 \quad \text{for all } 0 \neq v_0 \in V_{0,n},$$
$$\tag{7.27b}$$

$$\inf_{w \in W_n \text{ with } \|w\|_W = 1} \sup_{x \in V_n \text{ with } \|v\|_V = 1} |b(w, x)| \geq \beta > 0, \tag{7.27c}$$

where

$$V_{0,n} := \ker(B_n^*) = \{v \in V : b(y, v) = 0 \text{ for all } y \in W_n\} \subset V_n.$$

The uniform boundedness of C_n^{-1} is expressed by the fact that the positive numbers α, β in (7.27a,c) are independent of n (cf. §12.3.2 in [8, 10]).

The solvability of the undiscretised saddle-point problem is characterised by the same conditions (7.27a-c), but with $V_{0,n}$ and V_n replaced by $V_0 = \ker(B^*)$ and V. In the case of Stokes' problem (7.26), $a(\cdot, \cdot)$ is coercive on the whole space V, but for elasticity problems, one must exploit the fact that coercivity is only needed for the smaller subspace $V_0 \subset V$ (cf. Braess [2, §III.4 and §VI]).

7.5.8 Further Remarks

7.5.8.1 Perturbations of A_n and f_h

The determination of A_n and f_h requires the evaluation of integrals. This can be done exactly for A_n in cases of constant coefficients as in (7.1). However, for general f and general coefficients as in (7.2), the integral must be approximated by some quadrature thereby producing additional perturbations δA_n and δf_n. Whatever the origin of these perturbations, the natural requirements are

$$\|\delta A_n\|_{X_n^* \leftarrow X_n} \to 0, \quad \|\delta f_n\|_{X_n^*} \to 0 \quad \text{as } \mathbb{N}' \ni n \to \infty. \tag{7.28}$$

The following statement has a similar flavour as in Theorems 3.46 and 3.47.

Theorem 7.24. *Suppose that the Galerkin approximation, characterised by* $\{A_n, f_n\}_{n \in \mathbb{N}'}$, *is stable with* $\|A_n^{-1}\|_{X_n \leftarrow X_n^*} \leq C_{\text{stab}}$. *Then, under assumption* (7.28), *also the perturbed discretisation* $\{A_n + \delta A_n, f_n + \delta f_n\}_{n \in \mathbb{N}'}$ *is stable. The resulting solution* $\tilde{u}_n = u_n + \delta u_n$ (u_n: *unperturbed discrete solution) satisfies the asymptotic inequality*

$$\|\delta u_n\|_{X_n} \tag{7.29}$$
$$\leq \left[\|\delta A_n\|_{X_n^* \leftarrow X_n} C_{\text{stab}} \|f_n\|_{X_n^*} + \|\delta f_n\|_{X_n^*} \right] \left(C_{\text{stab}} + O(\|\delta A_n\|_{X_n^* \leftarrow X_n}) \right).$$

Proof. Write $A_n + \delta A_n$ as $A_n(I + A_n^{-1}\delta A_n)$ and note that

$$\|A_n^{-1}\delta A_n\|_{X_n \leftarrow X_n} \leq \|A_n^{-1}\|_{X_n \leftarrow X_n^*}\|\delta A_n\|_{X_n^* \leftarrow X_n} \leq C_{\text{stab}}\|\delta A_n\|_{X_n^* \leftarrow X_n} \to 0.$$

Define the subset $\mathbb{N}'' := \{n \in \mathbb{N}' : \|\delta A_n\|_{X_n^* \leftarrow X_n} \leq \frac{1}{2C_{\text{stab}}}\}$. Then, for $n \in \mathbb{N}''$, we have $\|A_n^{-1}\delta A_n\|_{X_n \leftarrow X_n} \leq 1/2$ and $\|(I + A_n^{-1}\delta A_n)^{-1}\|_{X_n \leftarrow X_n} \leq 2$ (cf. Lemma 5.8), so that

$$\|(A_n + \delta A_n)^{-1}\|_{X_n \leftarrow X_n^*} = \|(I + A_n^{-1}\delta A_n)^{-1}A_n^{-1}\|_{X_n \leftarrow X_n^*} \leq 2C_{\text{stab}} =: C'_{\text{stab}}.$$

This proves the stability of $A_n + \delta A_n$.

$(A_n + \delta A_n)^{-1} - A_n^{-1} = [(I + A_n^{-1}\delta A_n)^{-1} - I]A_n^{-1}$ can be estimated by

$$\|(A_n+\delta A_n)^{-1}-A_n^{-1}\|_{X_n \leftarrow X_n^*} \leq C_{\text{stab}}\|\delta A_n\|_{X_n^* \leftarrow X_n}(C_{\text{stab}}+O(\|\delta A_n\|_{X_n^* \leftarrow X_n})),$$

while $\|(A_n+\delta A_n)^{-1}\|_{X_n \leftarrow X_n^*} \leq (C_{\text{stab}}+\|\delta A_n\|_{X_n^* \leftarrow X_n})$. Therefore, we conclude that

$$\delta u_n = \tilde{u}_n - u_n = (A_n + \delta A_n)^{-1}(f_n + \delta f_n) - A_n^{-1}f_n$$
$$= [(A_n + \delta A_n)^{-1} - A_n^{-1}]f_n + (A_n + \delta A_n)^{-1}\delta f_n$$

can be estimated as in (7.29). $\qquad\square$

As an illustration, we discuss the case of quadrature mentioned above. The non-vanishing entries a_{ij} of the finite element matrix A_n are of the size $O(h^{-2+d/2})$ when we assume a quasi-uniform discretisation with maximal grid size h in the d-dimensional domain $\Omega \subset \mathbb{R}^d$. A general quadrature method of the second order leads to an absolute error of size $O(h^{d/2})$ and to the spectral norm $\|\delta A_n\|_2 = O(h^{d/2})$. Since $\|\cdot\|_{X_n^* \leftarrow X_n} \leq \|\cdot\|_2$, the first inequality in (7.28) is satisfied.

Another error treatment follows the idea of backward analysis. Let $a(u, v) = \int_\Omega \langle \nabla v, c(x)\nabla u \rangle \, dx$ be the bilinear form with a (piecewise) smooth coefficient c. For piecewise linear finite elements b_i, the integrals $\int_\Delta \langle \nabla b_i, c(x)\nabla b_j \rangle \, dx$ are const $\cdot \int_\Delta c(x) dx$ (Δ: triangle). Let c_Δ be the quadrature result of $\int_\Delta c(x) dx$ and define a new boundary value problem with piecewise constant coefficients $\tilde{c}(x) := c_\Delta / \int_\Delta dx$ for $x \in \Delta$. Then the finite element matrix $A_n + \delta A_n$ is the exact matrix for the new bilinear form $\tilde{a}(\cdot, \cdot)$.

Another formulation of the total error including (7.29) is given by the *first Strang lemma* (cf. Braess [2, Part III, §1]): for coercive bilinear forms $a(\cdot, \cdot)$ and $a_n(\cdot, \cdot)$ and right-hand sides f and f_n let u and u_n be the respective solutions of

$$a(u, v) = f \text{ for all } v \in X, \qquad a_n(u_n, v_n) = f_n \text{ for all } v_n \in X_n \subset X.$$

Then

$$\|u - u_n\|_X \leq C\left\{ \inf_{v_n \in X_n} \|u - v_n\|_X + \sup_{w_n \in X_n} \frac{|a(u_n, w_n) - a_n(u_n, w_n)|}{\|w_n\|_X} \right.$$
$$\left. + \sup_{w_n \in X_n} \frac{|(f - f_n)(w_n)|}{\|w_n\|_X} \right\}.$$

7.5.8.2 Nonconforming Methods

The standard conforming finite element methods are characterised by the inclusion $X_n \subset X$, which allows us to evaluate $a(u_n, v_n)$ for $u_n, v_n \in X_n$. This is also called *internal* approximation. An *external* approximation uses a sequence of spaces X_n containing also elements outside of X; i.e.,

$$X_n \not\subset X.$$

The variational formulation of the nonconforming Galerkin method is of the form

$$\text{find } u_n \in X_n \text{ with } \quad a_n(u_n, v_n) = f_n(v_n) \quad \text{for all } v_n \in X_n, \tag{7.30}$$

where the subscript of a_n indicates that the bilinear form depends on n. The associated operator $A_n : X_n \to X_n'$ is defined by

$$(A_n u_n)(v_n) = a_n(u_n, v_n) \quad \text{for all } u_n, v_n \in X_n.$$

Therefore, the variational formulation (7.30) is equivalent to

$$A_n u_n = f_n. \tag{7.31}$$

Above we defined a vector space X_n, but we have not yet defined a norm. In general, this norm depends on n:

$$X_n \text{ normed by } \||\cdot\||_n.$$

As mentioned above, the bilinear form $a_n(\cdot, \cdot) : X_n \times X_n \to \mathbb{R}$ defines the operator $A_n : X_n \to X_n'$ via $a_n(x_n, \cdot) = \langle A_n x_n, \cdot \rangle_{X_n' \times X_n}$. *Stability* is characterised by the existence of the inverse A_n^{-1}—at least for an infinite subset $\mathbb{N}' \subset \mathbb{N}$—and the uniform estimate

$$C_{\text{stab}} := \sup_{n \in \mathbb{N}'} \left\| A_n^{-1} \right\|_{X_n' \leftarrow X_n} < \infty.$$

Since X_n is finite-dimensional, the latter statement is equivalent to the inf-sup condition

$$\inf_{0 \neq v_n \in V_n} \sup_{0 \neq w_n \in V_n} \frac{|a_n(v_n, w_n)|}{\||v_n\||_n \||w_n\||_n} \geq \frac{1}{C_{\text{stab}}} > 0. \tag{7.32}$$

Before we study the connection between $a_n(\cdot, \cdot)$ and $a(\cdot, \cdot)$, we consider the functionals f_n and f. The usual setting $a(u, v) = f(v)$ $(v \in X)$ describes the problem $Au = f$ with a functional $f \in X'$. The obvious discrete formulation

$$a_n(u_n, v_n) = f(v_n) \quad \text{for all } v_n \in X_n \tag{7.33}$$

is not quite sound, since f may be undefined for $v_n \in X_n \backslash X$ (note that $X_n \backslash X \neq \emptyset$ in the nonconforming case). There are two remedies: (a) Extend f continuously to f_n such that $f(v) = f_n(v)$ for $v \in X$. (b) Usually, we have a Gelfand triple

$$X \subset U = U' \subset X'$$

(e.g., with $U = L^2(\Omega)$, cf. §6.3.3 in [8, 10]). If $f \in U'$ and $X_n \subset U$, the functional f is well defined on X_n and the variational formulation (7.33) makes sense (i.e., we may choose $f_n := f$).

The restriction of f to a smaller subset $U' \subsetneqq X'$ implies that also the solution u belongs to a subspace strictly smaller than X. We denote this space by X_*. In the case of $f \in U'$, this is

$$X_* := \{A^{-1}f : f \in U'\} \subsetneqq X.$$

There is always the problem of how to compare the discrete solution $u_n \in X_n$ and the exact solution $u \in X$ if the spaces X_n and X are different. One possibility is to restrict $u \in X$ to X_n and to compare both in X_n (cf. (7.6a)). Another approach is to extend $u_n \in X_n$ into X (cf. (7.6b)). The latter case is trivial in the conforming Galerkin case, since we have the natural inclusion $X_n \subset X$. In the nonconforming case, neither X_n nor X contain both u and u_n. This problem leads us to the next construction: we define a common superspace

$$X_{*n} \supset X_* \text{ and } X_{*n} \supset X_n \text{ with norm } \interleave \cdot \interleave_{*n}.$$

Then the difference $u - u_n$ belongs to X_{*n} and convergence can be formulated by $\interleave u - u_n \interleave_{*n} \to 0$. We require that

$$\interleave \cdot \interleave_n \leq \bar{C}_* \interleave \cdot \interleave_{*n} \qquad \text{on } X_n \tag{7.34}$$

with a constant $\bar{C}_*$. Without loss of generality, we may scale $\interleave \cdot \interleave_{*n}$ such that $\bar{C}_* = 1$.

The next requirement is that the bilinear form $a_n(\cdot, \cdot)$ can be continuously extended from $X_n \times X_n$ onto $X_{*n} \times X_n$ such that

$$|a_n(v, w)| \leq C_a \interleave v \interleave_{*n} \interleave w \interleave_n \text{ for all } v \in X_{*n}, w \in X_n. \tag{7.35}$$

Remark 7.25. *Assume that $(V, \|\cdot\|_V)$ and $(W, \|\cdot\|_W)$ are Banach spaces with an intersection $V \cap W$ possibly larger than the zero space $\{0\}$. Then a canonical norm of the smallest common superspace $U := \operatorname{span}(V, W)$ is defined by $\|u\|_U := \inf\{\|v\|_V + \|w\|_W : u = v + w, v \in V, w \in W\}$.*

Corollary 7.26. *Assume that $\|\cdot\|_V$ and $\|\cdot\|_W$ are equivalent norms on $V \cap W$. Prove that $\|v\|_V \leq \|v\|_U \leq C \|v\|_V$ for all $v \in V$ and $\|w\|_W \leq \|w\|_U \leq C \|w\|_W$ for all $w \in W$, where $\|\cdot\|_U$ is the norm from Remark 7.25.*

Consistency is expressed by

$$a_n(u, v_n) = f(v_n) \qquad \text{for all } v_n \in X_n, \tag{7.36}$$

where u is the solution of $a(u, v) = f(v)$ for $v \in X$ (here, we assume that f belongs to X_n'). Then the error estimate (Strang's second lemma, cf. [18, §1.3])

$$\|\|u - u_n\|\|_n \leq \left(\bar{C}_* + C_a C_{\text{stab}}\right) \inf_{w_n \in X_n} \|\|u - w_n\|\|_{*n} + C_a C_{\text{stab}} \|\|f_n - f\|\|_n^*$$

holds, where $\|\| \cdot \|\|_n^*$ is the norm dual to $\|\| \cdot \|\|_n$.

Proof. Let u_n be the discrete solution, while $w_h \in X_n$ is arbitrary. From (7.32) we infer that

$$\|\|u_n - w_n\|\|_n \leq C_{\text{stab}} \sup_{v_n \in X_n \text{ with } \|\|v_n\|\|_n = 1} |a_n(u_n - w_n, v_n)|.$$

Split $a_n(u_n - w_n, v_n)$ into $a_n(u_n - u, v_n) + a_n(u - w_n, v_n)$. The first term can be reformulated in terms of (7.30) and (7.36): $a_n(u_n - u, v_n) = (f_n - f)(v_n)$; hence

$$|a_n(u_n - u, v_n)| \leq \|\|f_n - f\|\|_n^*$$

because $\|\|v_n\|\|_n = 1$. The second term is estimated by

$$|a_n(u - w_n, v_n)| \leq C_a \|\|u - w_n\|\|_{*n}$$

(cf. (7.35)). Together, we obtain

$$\|\|u_n - w_n\|\|_n \leq C_{\text{stab}} \left(\|\|f_n - f\|\|_n^* + C_a \|\|u - w_n\|\|_{*n} \right).$$

The triangle inequality yields

$$\begin{aligned}
\|\|u - w_n\|\|_n &\leq \|\|u - u_n\|\|_n + \|\|u_n - w_n\|\|_n \\
&\leq \bar{C}_* \|\|u - u_n\|\|_{*n} + \|\|u_n - w_n\|\|_n \\
&\leq \left(\bar{C}_* + C_a C_{\text{stab}}\right) \|\|u - w_n\|\|_{*n} + C_a C_{\text{stab}} \|\|f_n - f\|\|_n^*.
\end{aligned}$$

Taking the infimum over all w_n, we obtain the error estimate. $\square$

The consistency condition (7.36) is not always satisfied in practical examples as, e.g., Wilson's element (for a definition and analysis compare, e.g., Shi [22]). As a remedy, one has tried to formulate criteria that are easier to check, the so-called *patch tests*. Stummel [24, 25] analysed the patch tests and proved that they are neither necessary nor sufficient for stability.

7.5.8.3 Discontinuous Galerkin Method

A prominent example of a nonconforming finite element is the discontinuous Galerkin method (DG), which started in the field of hyperbolic problems and for elliptic problems with dominant convection. Now it has become a popular method also for standard elliptic problems (see [18], [3], [14], [21, p. 255]).

The spaces X_* and $X_{*n} \supset X_n \supset X_*$ are as follows. $X_* = H_0^1(\Omega)$ is the standard choice. Given any triangulation $\mathcal{T}_n$ of Ω, the *broken* H^1 space is defined by

$$X_{*n} := \{u|_T \in H^1(T) \text{ for all } \tau \in \mathcal{T}_n\} \text{ with } \|u\|_{X_n} := \sqrt{\sum_{\tau \in \mathcal{T}_n} \|u_n\|_{H^1(\tau)}^2}.$$

The remarkable fact is that no conditions are required concerning the connection of $u|_{T'}$ and $u|_{T''}$ for neighboured elements $T', T'' \in \mathcal{T}_n$; i.e., in general, functions from X_{*n} are discontinuous across the internal boundaries of $\mathcal{T}_n$. Obviously, $X_{*n} \supset X_* = H_0^1(\Omega)$ holds.

The DG finite element space is the subspace X_n of X_{*n}, where all $u|_T$ are, e.g., piecewise linearly affine. The advantage of the discontinuous Galerkin method becomes obvious when we want to choose different polynomial degrees for different elements. Hence, the computational overhead for an hp-method is much lower.

Starting from a strong solution of $-\Delta u = f$ in Ω with $u = 0$ on $\Gamma = \partial\Omega$, multiplication with a test function $v \in X_n$ and partial integration in each $T \in \mathcal{T}_n$ yield the variational form (7.36) with

$$\int_\Omega (-\Delta u)\, v_n \mathrm{d}x = \sum_{\tau \in \mathcal{T}_n} \int_\tau \langle \nabla u_n, \nabla v_n \rangle \, \mathrm{d}x - \sum_{E \in \mathcal{E}_n} \int_E \left\{ \frac{\partial u}{\partial n_E} \right\} [v_n]\, \mathrm{d}s$$

and $f_n(v) = \int_\Omega fv \mathrm{d}x$. Here, $\mathcal{E}_n$ is the set of all edges of $\tau \in \mathcal{T}_n$. Each edge $E \in \mathcal{E}_n$ is associated with a normal n_E (the sign of the direction does not matter). The curly bracket $\{\ldots\}$ is the average of the expression in the two elements containing E, while $[\ldots]$ is the difference (its sign depends on the direction of n_E). The right-hand side in the latter expression is no longer symmetric. Since $[u] = 0$ for $u \in X_*$, we may add further terms without changing the consistency property:

$$a_n(u_n, v_n) := \sum_{\tau \in \mathcal{T}_n} \int_\tau \langle \nabla u_n, \nabla v_n \rangle \, \mathrm{d}x \qquad (7.37)$$

$$- \sum_{E \in \mathcal{E}_n} \int_E \left\{ \frac{\partial u_n}{\partial n_E} \right\} [v_n] - \left\{ \frac{\partial v_n}{\partial n_E} \right\} [u_n]\, \mathrm{d}s$$

$$+ \eta \sum_{E \in \mathcal{E}_n} h_E^{-1} \int_E [u_n]\,[v_n]\, \mathrm{d}s.$$

The expression in the second line is antisymmetric. Therefore it vanishes for $u_n = v_n$, and

$$a_n(u_n, u_n) = \sum_{\tau \in \mathcal{T}_n} \|\nabla u_n\|_{L^2(\tau)}^2 + \eta \sum_{E \in \mathcal{E}_n} h_E^{-1} \int_E [u_n]^2\, \mathrm{d}s$$

proves coercivity of a_n.

Other variants of the discontinuous Galerkin methods use a symmetric bilinear form with the term $\sum_{E \in \mathcal{E}_n} \int_E \left\{ \frac{\partial u_n}{\partial n_E} \right\} [v_n] + \left\{ \frac{\partial v_n}{\partial n_E} \right\} [u_n]\, \mathrm{d}s$ in the second line of (7.37) (cf. [18, pp. 124ff]).

7.5.9 Consistency Versus Stability

7.5.9.1 Convection-Diffusion Equation

For ordinary differential equations, we mentioned stiff differential equations (cf. §5.5.7.1). In that case, standard stable schemes work in the limit $h \to 0$, but their results are useless unless $h \le h_0$, where h_0 may be a rather small number. A similar phenomenon can happen for the discretisation of elliptic problems.

The differential equation

$$-\Delta u + \langle c, \operatorname{grad} u \rangle = f \text{ in } \Omega \subset \mathbb{R}^2, \qquad u = 0 \text{ on } \Gamma = \partial\Omega, \tag{7.38}$$

is called singularly perturbed if $\|c\| \gg 1$. Because of the dominant convection term $\langle c, \operatorname{grad} u \rangle$, the problem can be considered as a perturbation of the hyperbolic problem $\langle c, \operatorname{grad} u \rangle = f$. Since the hyperbolic problem cannot be combined with boundary data $u = 0$ on all of Γ, the solution of (7.38) develops a boundary layer at the outflow part of the boundary.

A stable discretisation can be constructed as follows. The part $-\Delta u$ is discretised by the five-point formula (7.3). The treatment of $\langle c, \operatorname{grad} u \rangle$ depends on the signs of the components c_1, c_2 in $c = (c_1, c_2)$. Assume $c_1 > 0$. Then $c_1 \partial u / \partial x_1$ is replaced by the backward difference

$$\frac{c_1}{h} \left(u_{\nu,\mu} - u_{\nu-1,\mu} \right).$$

If $c_1 < 0$, the forward difference

$$\frac{c_1}{h} \left(u_{\nu+1,\mu} - u_{\nu,\mu} \right)$$

is used. The second term $c_2 \partial u / \partial x_2$ in $\langle c, \operatorname{grad} u \rangle$ is treated analogously. As a result, the discretisation matrix is an M-matrix (cf. Definition 7.2). This property helps to prove the stability of the discretisation. However, the consistency order is one because of the one-sided differences.

Second-order consistency can be obtained by replacing $c_i \partial u / \partial x_i$ with symmetric differences. The conflict with stability can be checked by the violation of the M-matrix sign conditions. If $c_1 > 0$, the coefficient of $u_{\nu+1,\mu}$ becomes

$$-\frac{1}{h^2} + \frac{c_1}{2h}$$

and fails to be non-positive unless $h \le 2/c_1$. Because of the assumption $\|c\| \gg 1$, the requirement $h \le 2/c_1$ limits the use of the second-order scheme to rather small step sizes.

This conflict between consistency and stability also occurs for finite element discretisation (the corresponding modifications are, e.g., streamline diffusion methods, cf. John–Maubach–Tobiska [12]).

7.5.9.2 Defect Correction Methods

A possible remedy is the use of the *defect correction scheme*. It is based on two different discretisations:

$$L_h u_h = f_h \tag{7.39a}$$

is a stable scheme, possibly of a low order of consistency, whereas the second scheme

$$L'_h u'_h = f'_h \tag{7.39b}$$

has an order of higher consistency. Scheme (7.39b) is not required to be stable. It may even be singular (i.e., a solution u'_h may not exist).

The method consists of two steps. First, the basic scheme (7.39a) is applied to obtain the starting value $u_h^{(0)}$:

$$u_h^{(0)} := L_h^{-1} f_h. \tag{7.40a}$$

Next the defect of $u_h^{(0)}$ with respect to the second scheme is computed:

$$d_h := L'_h u_h^{(0)} - f'_h. \tag{7.40b}$$

The correction using d_h is performed in the second and final step:

$$u_h^{(1)} := u_h^{(0)} - L_h^{-1} d_h. \tag{7.40c}$$

We shall show that $u_h^{(1)}$ has better consistency properties than $u_h^{(0)}$. Note that $u_h^{(1)}$ is neither a solution of (7.39a) nor of (7.39b). If (7.39a) has consistency order one, the defect correction by (7.40c) can increase the consistency order of $u_h^{(1)}$ only by one. Therefore, the steps (7.40b,c) must be repeated a few times, if the consistency order of (7.39b) is larger than two. However, it is essential that the number of iterations be a small finite number. Otherwise, the possible instability of (7.39b) would destroy the result.

For the proof, we assume the following properties:

$$u^*: \text{exact solution}; \quad u_h^* := R_h^X u^*,$$
$$\|L_h u_h^* - f_h\|_{H_h^{-1}} \le Ch, \quad \|L'_h u_h^* - f'_h\|_{H_h^{-1}} \le Ch^2, \tag{7.41}$$
$$\|L_h^{-1}\|_{H_h^1 \leftarrow H_h^{-1}} \le C, \quad \|L_h^{-1}\|_{H_h^2 \leftarrow L_h^2} \le C,$$
$$\|L_h - L'_h\|_{H_h^{-1} \leftarrow H_h^2} \le Ch.$$

Here L_h^2, H_h^1, H_h^2 are the discrete analogues of $L_h^2(\Omega)$, $H_h^1(\Omega)$, $H_h^2(\Omega)$ (with derivatives replaced by divided differences). H_h^{-1} is the dual space of H_h^1.

The last estimate of $L_h - L'_h$ in (7.41) can be obtained as follows. Split $L_h - L'_h$ into

$$L'_h \left(I - R^X_h P^X_h \right) - \left(R'^Y_h L - L'_h R^X_h \right) P^X_h + \left(R'^Y_h - R^Y_h \right) L P^X_h$$
$$+ \left(R^Y_h L - L_h R^X_h \right) P^X_h - L_h \left(I - R^X_h P^X_h \right)$$

and assume

$$\left\| I - R^X_h P^X_h \right\|_{H^1_h \leftarrow H^2_h}, \left\| R'^Y_h L - L'_h R^X_h \right\|_{H^{-1}_h \leftarrow H^2} \leq Ch,$$
$$\left\| R'^Y_h - R^Y_h \right\|_{H^{-1}_h \leftarrow L^2}, \left\| R^Y_h L - L_h R^X_h \right\|_{H^{-1}_h \leftarrow H^2} \leq Ch,$$
$$\left\| L'_h \right\|_{H^{-1}_h \leftarrow H^1_h}, \left\| L_h \right\|_{H^{-1}_h \leftarrow H^1_h}, \left\| L \right\|_{L^2 \leftarrow H^2}, \left\| P^X_h \right\|_{H^2 \leftarrow H^2_h} \leq C.$$

Lemma 7.27. *Under the suppositions (7.41), the result $u^{(1)}_h$ of the defect correction method satisfies the estimate $\| u^{(1)}_h - u^*_h \|_{H^1_h} \leq Ch^2$.*

Proof. We may rewrite $u^{(1)}_h - u^*_h$ as

$$u^{(1)}_h - u^*_h = u^{(0)}_h - u^*_h - L^{-1}_h \left[L'_h u^{(0)}_h - f'_h \right]$$
$$= \left[I - L^{-1}_h L'_h \right] \left[u^{(0)}_h - u^*_h \right] + L^{-1}_h \left[L'_h u^*_h - f'_h \right]$$
$$= L^{-1}_h \left[L_h - L'_h \right] \left[u^{(0)}_h - u^*_h \right] + L^{-1}_h \left[L'_h u^*_h - f'_h \right].$$

Consistency $\| L'_h u^*_h - f'_h \|_{H^{-1}_h} \leq Ch^2$ and stability $\| L^{-1}_h \|_{H^1_h \leftarrow H^{-1}_h} \leq C$ imply that the second term satisfies $\| L^{-1}_h \left[L'_h u^*_h - f'_h \right] \|_{H^1_h} \leq Ch^2$.

Similarly,

$$\left\| u^{(0)}_h - u^*_h \right\|_{H^2_h} = \left\| L^{-1}_h f_h - u^*_h \right\|_{H^2_h} \leq \left\| L^{-1}_h \right\|_{H^2_h \leftarrow L^2_h} \left\| L_h u^*_h - f_h \right\|_{L^2_h} \leq Ch$$

holds. The inequalities $\| L^{-1}_h \|_{H^1_h \leftarrow H^{-1}_h} \leq C$ and $\| L_h - L'_h \|_{H^2_h \leftarrow H^{-1}_h} \leq Ch$ yield the estimate

$$\| L^{-1}_h \left[L_h - L'_h \right] \left[u^{(0)}_h - u^*_h \right] \|_{H^1_h} \leq Ch^2$$

for the first term. Together, the assertion of the lemma follows. $\square$

The defect correction method was introduced by Stetter [23]. A further description can be found in Hackbusch [9, §14.2]. The particular case of diffusion with dominant convection is analysed by Hemker [11].

The application of the defect correction method is not restricted to elliptic problems. An application to a hyperbolic initial-value problem can be found in [5].

References

1. Aubin, J.P.: Behaviour of the error of the approximate solution of boundary value problems for linear operators by Galerkin's and finite difference methods. Ann. Scuola Norm. Sup. Pisa **21**, 599–637 (1967)

2. Braess, D.: Finite Elements: Theory, Fast Solvers, and Applications in Solid Mechanics, 3rd ed. Cambridge University Press, Cambridge (2007)
3. Cockburn, B., Karniadakis, G.E., Shu, C.W. (eds.): Discontinuous Galerkin Methods. Theory, Computation and Applications, *Lect. Notes Comput. Sci. Eng.*, Vol. 11. Springer, Berlin (2000).
4. Fréchet, M.: Sur les ensembles de fonctions et les opérations linéaires. C. R. Acad. Sci. Paris **144**, 1414–1416 (1907)
5. Hackbusch, W.: Bemerkungen zur iterativen Defektkorrektur und zu ihrer Kombination mit Mehrgitterverfahren. Rev. Roumaine Math. Pures Appl. **26**, 1319–1329 (1981)
6. Hackbusch, W.: On the regularity of difference schemes. Ark. Mat. **19**, 71–95 (1981)
7. Hackbusch, W.: Iterative Solution of Large Sparse Systems of Equations. Springer, New York (1994)
8. Hackbusch, W.: Elliptic differential equations. Theory and numerical treatment, *Springer Series in Computational Mathematics*, Vol. 18, 2nd ed. Springer, Berlin (2003)
9. Hackbusch, W.: Multi-grid methods and applications, *Springer Series in Computational Mathematics*, Vol. 4. Springer, Berlin (2003)
10. Hackbusch, W.: Theorie und Numerik elliptischer Differentialgleichungen, 3rd ed. (2005). www.mis.mpg.de/scicomp/Fulltext/EllDgl.ps
11. Hemker, P.W.: Mixed defect correction iteration for the solution of a singular perturbation problem. Comput. Suppl. **5**, 123–145 (1984)
12. John, V., Maubach, J.M., Tobiska, L.: Nonconforming streamline-diffusion-finite-element-methods for convection diffusion problems. Numer. Math. **78**, 165–188 (1997)
13. Jovanovič, B., Süli, E.: Analysis of finite difference schemes for linear partial differential equations with generalized solutions, *Springer Series in Computational Mathematics*, Vol. 45. Springer, London (2013)
14. Kanschat, G.: Discontinuous Galerkin methods for viscous incompressible flows. Advances in Numerical Mathematics. Deutscher Universitätsverlag, Wiesbaden (2007)
15. Nečas, J.: Sur la coercivité des formes sesquilinéares elliptiques. Rev. Roumaine Math. Pures Appl. **9**, 47–69 (1964)
16. Nitsche, J.: Ein Kriterium für die Quasi-Optimalität des Ritzschen Verfahrens. Numer. Math. **11**, 346–348 (1968)
17. Ostrowski, A.M.: Über die Determinanten mit überwiegender Hauptdiagonale. Comment. Math. Helv. **10**, 69–96 (1937)
18. Di Pietro, D.A., Ern, A.: Mathematical aspects of discontinuous Galerkin methods. Springer, Berlin (2011)
19. Riesz, F.: Sur une espèce de géométrie analytique des systèmes de fonctions sommables. C. R. Acad. Sci. Paris **144**, 1409–1411 (1907)
20. Riesz, F., Sz.-Nagy, B.: Functional Analysis. Dover Publ. Inc, New York (1990)
21. Roos, H.G., Stynes, M., Tobiska, L.: Numerical methods for singularly perturbed differential equations: Convection-diffusion and flow problems, *Springer Series in Computational Mathematics*, Vol. 24. Springer, Berlin (1996)
22. Shi, Z.C.: A convergence condition for the quadrilateral Wilson element. Numer. Math. **44**, 349–361 (1984)
23. Stetter, H.J.: The defect correction principle and discretization methods. Numer. Math. **29**, 425–443 (1978)
24. Stummel, F.: The generalized patch test. SIAM J. Numer. Anal. **16**, 449–471 (1979)
25. Stummel, F.: The limitations of the patch test. Int. J. Num. Meth. Engng. **15**, 177–188 (1980)

Chapter 8
Stability for Discretisations of Integral Equations

Since the paper of Nyström [7], integral equations are used to solve certain boundary value problems. Concerning the integral equation method and its discretisation, we refer to Sauter–Schwab [8] and Hsiao–Wendland [5]. The following considerations of stability hold as long as the integral kernel is weakly singular.

8.1 Integral Equations and Their Discretisations

8.1.1 Integral Equation, Banach Space

The *Fredholm integral equation of the second kind* is described by

$$\lambda f = g + Kf, \tag{8.1}$$

where $\lambda \in \mathbb{C}\backslash\{0\}$ and the function g are given together with an integral operator K defined by

$$(Kf)(x) = \int_D k(x,y)f(y)\mathrm{d}y \qquad \text{for all } x \in D. \tag{8.2}$$

If $\lambda = 0$, Eq. (8.1) is called *Fredholm's integral equation of the first kind* (cf. Fredholm [3]). The integration domain D is an (often compact) subset of $\mathbb{R}^n$. Other interesting applications lead to surface integrals; i.e., $D = \partial\Omega$ of some $\Omega \subset \mathbb{R}^n$.

The function g belongs to a Banach space X with norm $\|\cdot\|$. The solution f of (8.1) is also sought in X: $g, f \in X$. The Banach space X must be chosen such that

$$K \in \mathcal{L}(X,X)$$

holds; i.e., K is a bounded mapping from X into itself. An equivalent notation of problem (8.1) is[1]

$$(\lambda I - K)f = g.$$

---

[1] Equations $(\lambda A - K)f = g$ with a not necessarily bounded operator A are considered in [6].

W. Hackbusch, *The Concept of Stability in Numerical Mathematics*, Springer Series in Computational Mathematics 45, DOI 10.1007/978-3-642-39386-0_8, © Springer-Verlag Berlin Heidelberg 2014

If $\lambda I - K : X \to X$ is bijective, also the inverse $(\lambda I - K)^{-1}$ belongs to $\mathcal{L}(X, X)$ (this is a consequence of the *open mapping theorem*; cf. [10, §II.5]) and a formal description of the solution is

$$f = (\lambda I - K)^{-1} g.$$

The operator norm $\|\cdot\|_{X \leftarrow X}$ is also denoted by $\|\cdot\|$.

The standard choices for X are $X = C(D)$ or $X = L^2(D)$ with the norms $\|f\| = \sup_{x \in D} |f(x)|$ or $\|f\| = \sqrt{\int_D |f(x)|^2 dx}$, respectively.

In many cases, the operator K is not only bounded, but also compact (cf. §4.9).

Lemma 8.1. *(a) If $k \in L^2(D \times D)$, then $K : X \to X$ is compact for $X = L^2(D)$.* *(b) If D is compact and $k \in C(D \times D)$, then $K : X \to X$ is compact for $X = C(D)$.*

Proof. The proof will be postponed to §8.1.2.1.   □

A more general criterion in the case of $X = C(D)$ is the following.

Theorem 8.2. *Suppose that D is compact and that k satisfies*

$$\int_D |k(x, y)| \, dy < \infty, \quad \lim_{\xi \to x} \int_D |k(\xi, y) - k(x, y)| \, dy = 0 \quad \text{for all } x \in D. \quad (8.3)$$

Then K is compact for $X = C(D)$.

Proof. (i) Precompactness of $M := \{Kf : f \in C(D), \|f\| \leq 1\}$ follows by Arzelà–Ascoli's theorem (cf. Theorem 3.51). Part (iii) will show uniform boundedness of M, while Part (iv) will yield equicontinuity.

(ii) We introduce the following auxiliary functions:

$$\varphi(x) := \int_D |k(x, y)| \, dy, \qquad \Phi(\xi, x) := \int_D |k(\xi, y) - k(x, y)| \, dy.$$

Claim: $\varphi \in C(D)$ is bounded, and $\Phi \in C(D \times D)$ is uniformly continuous. From $|\varphi(\xi) - \varphi(x)| = |\int_D (|k(\xi, y)| - |k(x, y)|) \, dy| \leq \int_D |k(\xi, y) - k(x, y)| \, dy = \Phi(\xi, x)$ and (8.3), it follows that φ is continuous. A continuous function on a compact set is bounded. An analogous estimate shows $|\Phi(\xi, x) - \Phi(\xi, x')| \leq \Phi(x, x')$ and $|\Phi(\xi, x) - \Phi(\xi', x)| \leq \Phi(\xi, \xi')$. Again by (8.3), Φ is continuous on the compact set $D \times D$; hence, uniformly continuous.

(iii) Each $g = Kf \in M$ is bounded by $\|g\| \leq \|\varphi\|$, since

$$|g(x)| = \left| \int_D k(x, y) f(y) dy \right| \leq \int_D |k(x, y)| \, |f(y)| \, dy \leq \varphi(x) \|f\| \leq \varphi(x) \leq \|\varphi\|.$$

Hence M is uniformly bounded.

(iv) Fix $\varepsilon > 0$. As Φ is uniformly continuous (cf. (ii)), there is some $\delta > 0$ such that $\Phi(\xi, x) = |\Phi(\xi, x) - \Phi(x, x)| \leq \varepsilon$ for all $\xi, x \in D$ with $|\xi - x| \leq \delta$. For any $g = Kf \in M$ we have

$$|g(\xi) - g(x)| = \left| \int_D [k(\xi, y) - k(x, y)] f(y) \mathrm{d}y \right| \leq \int_D |k(\xi, y) - k(x, y)| \overbrace{|f(y)|}^{\leq 1} \mathrm{d}y$$
$$\leq \Phi(\xi, x) \leq \varepsilon \qquad \text{for } |\xi - x| \leq \delta,$$

proving that all functions $g \in M$ are equicontinuous. $\quad\square$

We mention that the conditions from (8.3) are also necessary, provided that $k(x, \cdot) \in L^1(D)$ for all $x \in D$.

8.1.2 Discretisations

There are various discretisation techniques leading to a sequence of 'discrete' integral operators $K_n \in \mathcal{L}(X, X)$, $n \in \mathbb{N}$, with the property that the range of K_n is n-dimensional. The corresponding 'discrete' integral equation is

$$\lambda f_n = g + K_n f_n \tag{8.4a}$$

or
$$\lambda f_n = g_n + K_n f_n, \tag{8.4b}$$

where in the latter case
$$g_n \to g \tag{8.4c}$$

is assumed.

Because the right-hand sides in (8.4a,b) are contained in an at most $(n + 1)$-dimensional subspace, $\lambda \neq 0$ implies that f_n belongs to this subspace. This leads to a system of finitely many linear equations.

8.1.2.1 Kernel Approximation

The *kernel approximation* approximates the kernel function $k(\cdot, \cdot)$ by a so-called degenerate kernel

$$k_n(x, y) = \sum_{j=1}^{n} a_j(x) b_j(y) \quad \text{for } x, y \in D, \tag{8.5}$$

leading to

$$K_n f = \int_D k_n(\cdot, y) f(y) \mathrm{d}y = \sum_{j=1}^{n} a_j(\cdot) \int_D b_j(y) f(y) \mathrm{d}y.$$

Exercise 8.3. *Let $X = C(D)$ or $X = L^2(D)$. Suppose that $\|k_n - k\|_{C(D \times D)} \to 0$ or $\|k_n - k\|_{L^2(D \times D)} \to 0$, respectively. Prove the corresponding operator norm convergence $\|K - K_n\| \to 0$.*

The suppositions of Exercise 8.3 imply even compactness of K.

Remark 8.4. *Suppose that K_n has a finite-dimensional range. Then $\|K-K_n\|\to 0$ implies that $K, K_n : X \to X$ are compact operators.*

Proof. We recall two properties of compact operators: (i) finite dimension range implies compactness, (ii) a limit of compact operators is compact. From (i) we infer that K_n is compact, while (ii) proves that $K = \lim K_n$ is compact. $\square$

Proof of Lemma 8.1. It is sufficient to verify the supposition of Exercise 8.3. For Part (b) involving $X = C(D)$, use Weierstrass' approximation theorem. Since $D \times D$ is compact, there is a sequence of polynomials $P_n(x, y)$ of degree $\leq n - 1$ such that $\|P_n - k\|_{C(D \times D)} \to 0$. Since $k_n := P_n$ has a representation (8.5) with $a_j(x) = x^j$ and a polynomial $b_j(y)$ of degree $\leq n - 1$, Exercise 8.3 can be applied.

Proof. In the Hilbert case of $X = L^2(D)$, K possesses an infinite singular value decomposition[2]

$$K = \sum_{j=1}^{\infty} \sigma_j a_j b_j^* \quad \text{with orthonormal } \{a_j\}, \{b_j\} \subset X,$$

where the singular values satisfy $\sigma_1 \geq \sigma_2 \geq \ldots \geq 0$, $\sigma_1 = \|K\|$, and

$$\sum_{j=1}^{\infty} \sigma_j^2 = \|k\|_{L^2(D \times D)}^2 < \infty$$

(cf. [9], [4, §4.4.3]). The operator $K_n = \sum_{j=1}^{n} \sigma_j a_j b_j^*$ has the kernel

$$k_n(x, y) = \sum_{j=1}^{n} \sigma_j a_j(x) b_j(y)$$

and satisfies $\|K - K_n\|^2 = \sum_{j=n+1}^{\infty} \sigma_j^2 \to 0$. $\square$

8.1.2.2 Projection Method

The *projection method* is characterised by a sequence of subspaces $X_n \subset X$ and projections $\Pi_n \in \mathcal{L}(X, X)$ with

$$X_n = \text{range}(\Pi_n) \quad \text{with } \dim(X_n) = n. \tag{8.6}$$

Either we construct a sequence of projections $\{\Pi_n\}$ and define X_n by (8.6), or we build a sequence of subspaces $\{X_n\}$ and define Π_n as a projection onto X_n. Setting

$$K_n := \Pi_n K \quad \text{and} \quad g_n := \Pi_n g,$$

we obtain the discrete problem by (8.4b): $\lambda f_n = g_n + K_n f_n$. Note that the solution belongs to X_n, provided that $\lambda \neq 0$ and the solution exists.

---

[2] b_j' denotes the corresponding functional $b_j'(\varphi) := \int_D b_j(y)\varphi(y)dy$; formally, $b_j' := J b_j$ is obtained by the Riesz isomorphism explained in Remark 7.18.

There are two standard choices of Π_n.

(1) Case $X = C(D)$. Define an interpolation at the points $\{\xi_{k,n} : 1 \leq k \leq n\} \subset D$ by

$$\Pi_n \varphi := \sum_{k=1}^{n} \varphi(\xi_{k,n}) L_{k,n}$$

with Lagrange functions $L_{k,n} \in X$ (cf. page 48). Since $f_n \in X_n$, it suffices to know the function values $f_{n,k} := f_n(\xi_{k,n})$. The generation of the system of linear equations requires computing the integrals

$$(KL_{k,n})(\xi_{k,n}) = \int_D k(\xi_{k,n}, y) L_{k,n}(y) \mathrm{d}y.$$

(2) Case $X = L^2(D)$. The Hilbert space structure of X allows us to define Π_n as the orthogonal projection onto X_n, where $X_n = \mathrm{span}\{\varphi_{k,n} : 1 \leq k \leq n\}$ is some Galerkin subspace of dimension n. The generation of the linear system requires computing

$$\int_D \int_D \varphi_{k,n}(x) k(x, y) \varphi_{k,n}(y) \mathrm{d}x \mathrm{d}y.$$

8.1.2.3 Nyström Method

The integral $\int_D \ldots \mathrm{d}y$ suggests applying a quadrature method

$$\int_D \varphi(y) \mathrm{d}y \approx Q_n(\varphi) = \sum_{j=1}^{n} a_{j,n} \varphi(\xi_{j,n}).$$

This leads to the Nyström discretisation

$$\lambda f_n(x) = g(x) + \sum_{k=1}^{n} a_{k,n} k(x, \xi_{k,n}) f_n(\xi_{k,n}) \quad \text{for all } x \in D \qquad (8.7)$$

(cf. Nyström [7]). Restricting the latter equation to $x \in \{\xi_{k,n} : 1 \leq k \leq n\}$, we obtain a linear system for $f_n(\xi_{k,n})$, $1 \leq k \leq n$. If this system has a unique solution, the values $f_n(\xi_{k,n})$ inserted into the right-hand side of (8.7) determine $f_n(x)$ for all $x \in D$, provided that $\lambda \neq 0$. The extension of $f_n(\xi_{k,n})$ to $f_n \in C(D)$ is also called Nyström interpolation. Because of this interpolation, the Nyström discretisation can again be written as $\lambda f_n = g + K_n f_n$ with

$$(K_n \varphi)(x) = \sum_{k=1}^{n} a_{k,n} k(x, \xi_{k,n}) \varphi(\xi_{k,n}). \qquad (8.8)$$

Since we need point evaluations, the Banach space $X = C(D)$ must be used.

8.2 Stability Theory

8.2.1 Consistency

Let $K, K_n \in \mathcal{L}(X, X)$ be operators acting in a Banach space X.

Definition 8.5 (consistency). $\{K_n\}$ is called consistent with respect to K if[3]

$$K_n \varphi \to K \varphi \qquad \text{for all } \varphi \in X. \tag{8.9}$$

$\{K_n\}$ is called consistent if some $K \in \mathcal{L}(X, X)$ exists satisfying (8.9).

Exercise 8.6. (a) Prove that $\{K_n\}$ is consistent with respect to K if and only if
(i) $K_n \varphi \to K \varphi$ for all $\varphi \in M \subset X$ and some M dense in X, and
(ii) $\sup_n \|K_n\| < \infty$.
   (b) Assuming that $K_n \varphi$ is a Cauchy sequence and $\sup_n \|K_n\| < \infty$, define K by
$K \varphi := \lim K_n \varphi$ and prove that $K \in \mathcal{L}(X, X)$; i.e., $\{K_n\}$ is consistent with respect
to K.
   (c) Operator norm convergence $\|K - K_n\| \to 0$ is sufficient for consistency.

8.2.2 Stability

Stability refers to the value $\lambda \neq 0$ from problem (8.1). This λ is assumed to be fixed.
Otherwise, one has to use the term 'stable with respect to λ'.

Definition 8.7 (stability). $\{K_n\}$ is called stable if there exist some $n_0 \in \mathbb{N}$ and
$C_{\text{stab}} \in \mathbb{R}$ such that[4]

$$\|(\lambda I - K_n)^{-1}\| \le C_{\text{stab}} \qquad \text{for all } n \ge n_0.$$

If $(\lambda I - K)^{-1} \in \mathcal{L}(X, X)$, the inverse exists also for perturbations of K.

Remark 8.8. Suppose $(\lambda I - K)^{-1} \in \mathcal{L}(X, X)$ and

$$\|K - K_n\| < 1/\|(\lambda I - K)^{-1}\|.$$

Then

$$\|(\lambda I - K_n)^{-1}\| \le \frac{\|(\lambda I - K)^{-1}\|}{1 - \|(\lambda I - K)^{-1}\| \|K - K_n\|}.$$

Proof. Apply Lemma 5.8 with $T := \lambda I - K$ and $S := \lambda I - K_n$. $\square$

---

[3] In Definition 4.5, consistency involves $K_n \varphi \to K \varphi$ only for φ from a dense subset $X_0 \subset X$.
Then the full statement (8.9) could be obtained from stability: $\sup_n \|K_n\| < \infty$. Here, stability
will be defined differently. This is the reason to define consistency as in (8.9).
[4] $\|(\lambda I - K)^{-1}\| \le C_{\text{stab}}$ is the short notation for '$\lambda I - K \in L(X, X)$ is bijective and the
inverse satisfies $\|(\lambda I - K)^{-1}\| \le C_{\text{stab}}$'.

A simple consequence is the following stability result.

Remark 8.9. *If* $(\lambda I - K)^{-1} \in \mathcal{L}(X, X)$ *and* $\|K - K_n\| \to 0$, *then* $\{K_n\}$ *is stable.*

The roles of K and K_n can be interchanged.

Exercise 8.10. *If* $\|K - K_n\| < 1/\|(\lambda I - K_n)^{-1}\|$ *holds for all* $n \geq n_0$, *then the inverse* $(\lambda I - K)^{-1} \in \mathcal{L}(X, X)$ *exists. The assumption is, in particular, valid, if* $\|K - K_n\| \to 0$ *and* $\{K_n\}$ *is stable.*

8.2.3 Convergence

Definition 8.11 (convergence). $\{K_n\}$ *is called convergent if there is some* $n_0 \in \mathbb{N}$ *such that*

(i) $\lambda f_n = g + K_n f_n$ *is solvable for all* $g \in X$ *and* $n \geq n_0$, *and*

(ii) the limit $\lim_n f_n$ *exists in* X.

The next remark shows that $f := \lim_n f_n$ satisfies the continuous problem. The existence of $(\lambda I - K)^{-1}$ is left open, but will follow later from Theorem 8.16.

Remark 8.12. *If* $\{K_n\}$ *is consistent and convergent, then* $f := \lim_n f_n$ *satisfies* $\lambda f = g + K f$ *(cf. (8.1)) and the operator* $\lambda I - K$ *is surjective.*

Proof. (i) By Exercise 8.6a, $C := \sup_n \|K_n\| < \infty$ holds. The solutions f_n exist for $n \geq n_0$ and define some $f := \lim_n f_n$. Consistency shows that

$$\|K_n f_n - Kf\| = \|K_n (f_n - f) - (K - K_n) f\|$$
$$\leq C \|f_n - f\| + \|(K - K_n) f\| \to 0;$$

hence the limit process $n \to \infty$ in $\lambda f_n = g + K_n f_n$ yields $\lambda f = g + Kf$.

(ii) Since for all $g \in X$, (i) yields a solution of $(\lambda I - K) f = g$, surjectivity follows. $\square$

A particular result of the Riesz–Schauder theory (cf. Yosida [10, §X.5]) is the following result.

Lemma 8.13. *Suppose that* $\lambda \neq 0$, $K \in \mathcal{L}(X, X)$ *compact. Then* $\lambda I - K$ *is injective if and only if* $\lambda I - K$ *is surjective.*

Exercise 8.14. *Let* K *be compact and* $\{K_n\}$ *be consistent and convergent. Prove* $(\lambda I - K)^{-1} \in \mathcal{L}(X, X)$.

Lemma 8.15. *If* $\{K_n\}$ *is stable and consistent, then the operator* $\lambda I - K$ *is injective.*

Proof. Injectivity follows from $\|(\lambda I - K)\varphi\| \geq \eta\|\varphi\|$ for some $\eta > 0$ and all $\varphi \in X$. For an indirect proof assume that there is a sequence $\varphi_n \in X$ with

$$\|\varphi_n\| = 1 \text{ and } \|\psi_n\| \leq 1/n \quad \text{for } \psi_n := (\lambda I - K)\varphi_n.$$

By consistency $K_m\varphi_n \to K\varphi_n$ for $m \to \infty$ and fixed n. Hence there is some $m_n \in \mathbb{N}$ such that $\|K_m\varphi_n - K\varphi_n\| \leq 1/n$ for $m \geq m_n$. We estimate the norm of $\zeta_{n,m} := (\lambda I - K_m)\varphi_n = \psi_n - (K_m\varphi_n - K\varphi_n)$ by $2/n$ for $m \geq m_n$. Stability and the representation $\varphi_n = (\lambda I - K_m)^{-1}\zeta_{n,m}$ yields $1 = \|\varphi_n\| \leq C \cdot 2/n$. The contradiction follows for $n > 2C$. $\square$

8.2.4 Equivalence

Theorem 8.16 (stability theorem). *(a) Convergence implies stability.*
(b) Convergence and consistency imply stability and $(\lambda I - K)^{-1} \in \mathcal{L}(X, X)$.

Proof. Part (a) follows again by Banach–Steinhaus (Corollary 3.39).

In the case of Part (b), Remark 8.12 states that $\lambda I - K$ is surjective. Since, by Part (a) stability holds, Lemma 8.15 proves that $\lambda I - K$ is also injective. Together, $\lambda I - K$ is bijective, which proves $(\lambda I - K)^{-1} \in \mathcal{L}(X, X)$. $\square$

Theorem 8.17 (convergence theorem). *Suppose consistency, stability, and either*
(i) $\lambda I - K$ surjective or (ii) K compact.
(a) Then $\{K_n\}$ is convergent and

$$f_n = (\lambda I - K_n)^{-1} g \to f = (\lambda I - K)^{-1} g \quad \text{for all } g \in X.$$

(b) If $g_n \to g$, then $f_n = (\lambda I - K_n)^{-1} g_n \to f = (\lambda I - K)^{-1} g$.

Proof. Lemma 8.15 ensures injectivity of $\lambda I - K$. Together with assumption (i), $(\lambda I - K)^{-1} \in \mathcal{L}(X, X)$ is proved. In the case of (ii), we conclude that injectivity of $\lambda I - K$ implies surjectivity (cf. Lemma 8.13), if $\lambda \neq 0$ and K compact. This allows us to define $f := (\lambda I - K)^{-1} g$. Set

$$d_n := \lambda f - K_n f - g_n,$$

where either $g_n := g$ (case of (8.4a)) or $g_n \to g$. Consistency and $g_n \to g$ show that $d_n \to 0$. Subtraction of $f_n := (\lambda I - K_n)^{-1} g_n$ implies that $(\lambda I - K_n)(f - f_n) = d_n$ and

$$f - f_n = (\lambda I - K_n)^{-1} d_n$$

for $n \geq n_0$. Stability yields the estimate $\|f - f_n\| \leq C\|d_n\| \to 0$; i.e., $f_n \to f$. $\square$

Combining the previous theorems, we obtain the next one.

Theorem 8.18 (equivalence theorem). *Suppose consistency and one of the conditions (i) or (ii) from Theorem 8.17. Then stability and convergence are equivalent. Furthermore, $g_n \to g$ implies*

$$f_n = (\lambda I - K_n)^{-1} g_n \to f = (\lambda I - K)^{-1} g.$$

Remark 8.19. *The suppositions $(\lambda I - K)^{-1} \in \mathcal{L}(X, X)$ and $\|K - K_n\| \to 0$ imply consistency, stability, and convergence.*

Proof. Exercise 8.6c shows consistency. Remark 8.9 yields stability. Finally, convergence is ensured by Theorem 8.17a. □

For the numerical solution of a linear problem $Ax = b$, the condition of A is important:

$$\mathrm{cond}(A) = \begin{cases} \|A\| \|A^{-1}\| & \text{if } A, A^{-1} \in \mathcal{L}(X, X), \\ \infty & \text{otherwise.} \end{cases} \tag{8.10}$$

Exercise 8.20. *If $\{K_n\}$ is consistent and convergent, then there is some $n_0 \in \mathbb{N}$ such that*

$$\sup_{n \geq n_0} \mathrm{cond}(\lambda I - K_n) < \infty.$$

8.3 Projection Methods

We recall that a projection method is characterised by the sequences $\{\Pi_n\}$ and $\{X_n = \mathrm{range}(\Pi_n)\}$. The approximation of X by $\{X_n\}$ is described by the condition

$$\mathrm{dist}(x, X_n) := \inf\{\|x - y\| : y \in X_n\} \to 0 \qquad \text{for all } x \in X. \tag{8.11}$$

Definition 8.21. *A sequence $\{\Pi_n\}$ is called convergent if $\Pi_n x = x$ for all $x \in X$.*

Exercise 8.22. *Prove that a convergent sequence $\{\Pi_n\}$ generates subspaces X_n satisfying condition (8.11).*

The terms consistency and stability can also be applied to $\{\Pi_n\}$.

Exercise 8.23. *Define:*
(i) $\{\Pi_n\}$ is consistent if there is a dense subset $M \subset X$ such that $\Pi_n x = x$ for all $x \in M$;
(ii) $\{\Pi_n\}$ is stable if $\sup_n \|\Pi_n\| < \infty$.
Prove that convergence (see Definition 8.21) is equivalent to consistency and stability.

8.4 Stability Theory for Nyström's Method

We recall that only the Banach space $X = C(D)$ (or a Banach space with even stronger topology) makes sense. Therefore, throughout this section $X = C(D)$ is chosen.

Lemma 8.24. *Assume that D is compact and $k \in C(D \times D)$. Then $\|K - K_n\| \geq \|K\|$.*

Proof. The operator norm $\|K\|$ can be shown to be equal to $\sup_{x \in D} \int_D |k(x,y)| \, dy$. Since D is compact, the supremum is a minimum; i.e., there is some $\xi \in D$ with $\|K\| = \int_D |k(\xi, y)| \, dy$. For any $\varepsilon > 0$, one can construct a function $\varphi_\varepsilon \in X = C(D)$ such that $\|\varphi_\varepsilon\| = 1$, $\|K\varphi_\varepsilon\| \geq \int_D |k(\xi, y)| \, dy - \varepsilon$, and in addition, $\varphi_\varepsilon(\xi_{k,n}) = 0$ for $1 \leq k \leq n$. The latter property implies $K_n \varphi_\varepsilon = 0$, so that $\|K - K_n\| \geq \|(K - K_n)\varphi_\varepsilon\| = \|K\varphi_\varepsilon\| \geq \|K\| - \varepsilon$. As $\varepsilon > 0$ is arbitrary, the assertion follows. $\square$

The statement of the lemma shows that we cannot use the argument that the operator norm convergence $\|K - K_n\| \to 0$ proves the desired properties.

It will turn out that instead of $K - K_n$, the products $(K - K_n)K$ and $(K - K_n)K_n$ may still converge to zero. The next *theorem of Brakhage* [2] proves the main step. Here we use the operators $S, T \in \mathcal{L}(X, X)$ which replace K and K_n (for a fixed n). In this theorem, X may be any Banach space.

Theorem 8.25. *Let X be a Banach space and $\lambda \neq 0$. Suppose that the operators $S, T, (\lambda I - S)^{-1}$ belong to $\mathcal{L}(X, X)$, and that T is compact. Under the condition*

$$\|(T - S)T\| < |\lambda| / \|(\lambda I - S)^{-1}\|, \tag{8.12a}$$

also $(\lambda I - T)^{-1} \in \mathcal{L}(X, X)$ exists and satisfies

$$\|(\lambda I - T)^{-1}\| \leq \frac{1 + \|(\lambda I - S)^{-1}\| \, \|T\|}{|\lambda| - \|(\lambda I - S)^{-1}\| \, \|(T - S)T\|}. \tag{8.12b}$$

The solutions of $(\lambda I - S) f_S = g$ and $(\lambda I - T) f_T = g$ differ by

$$\|f_S - f_T\| \leq \|(\lambda I - S)^{-1}\| \frac{\|(T - S)T\| \, \|f_S\| + \|(T - S)g\|}{|\lambda| - \|(\lambda I - S)^{-1}\| \, \|(T - S)T\|}, \tag{8.12c}$$

$$\|f_S - f_T\| \leq \|(\lambda I - T)^{-1}\| \, \|(T - S)f_S\|. \tag{8.12d}$$

Proof. (i) If $(\lambda I - T)^{-1} \in \mathcal{L}(X, X)$, the identity $I = \frac{1}{\lambda}[(\lambda I - T) + T]$ leads to $(\lambda I - T)^{-1} = \frac{1}{\lambda}[I + (\lambda I - T)^{-1} T]$. We replace $(\lambda I - T)^{-1}$ on the right-hand side by the inverse $(\lambda I - S)^{-1}$, whose existence is assumed, and define

$$A := \frac{1}{\lambda}[I + (\lambda I - S)^{-1} T].$$

$B := A(\lambda I - T)$ should approximate the identity:

$$B = \tfrac{1}{\lambda}[I + (\lambda I - S)^{-1} T](\lambda I - T) = I - \tfrac{1}{\lambda}[T - (\lambda I - S)^{-1} T(\lambda I - T)]$$
$$= I - \tfrac{1}{\lambda}(\lambda I - S)^{-1}[(\lambda I - S)T - T(\lambda I - T)]$$
$$= I - \tfrac{1}{\lambda}(\lambda I - S)^{-1}(T - S)T.$$

By assumption,

$$\left\| \tfrac{1}{\lambda}(\lambda I - S)^{-1}(T - S)T \right\| \leq \| (\lambda I - S)^{-1} \| \, \| (T - S)T \| / |\lambda| < 1$$

holds and proves the existence of B^{-1}.

(ii) Since $B = A(\lambda I - T)$ is injective (even bijective), also $\lambda I - T$ must be injective. Lemma 8.13 states that $\lambda I - T$ is surjective; hence, $\lambda I - T$ is bijective and $(\lambda I - T)^{-1} \in \mathcal{L}(X, X)$ exists. In particular, $B = A(\lambda I - T)$ implies $(\lambda I - T)^{-1} = B^{-1}A$:

$$(\lambda I - T)^{-1} = \left[\lambda I - (\lambda I - S)^{-1}(T - S)T \right]^{-1}[I + (\lambda I - S)^{-1} T].$$

The inverse on the right-hand side can be estimated by

$$\left\| \left[\lambda I - (\lambda I - S)^{-1}(T - S)T \right]^{-1} \right\| \leq \frac{1}{|\lambda| - \| (\lambda I - S)^{-1} \| \, \| (T - S)T \|}$$

(cf. Lemma 5.8). Together with $\| I + (\lambda I - S)^{-1} T \| \leq 1 + \| (\lambda I - S)^{-1} \| \, \| T \|$, we obtain inequality (8.12b).

(iii) Subtracting $(\lambda I - T) f_T = g$ from $(\lambda I - S) f_S = g$, we obtain the expression $\lambda(f_S - f_T) = S f_S - T f_T = T(f_S - f_T) + (S - T)f_S$ and

$$f_S - f_T = (\lambda I - T)^{-1}(S - T)f_S.$$

This proves inequality (8.12c). Interchanging S and T, we arrive at $f_T - f_S = (\lambda I - S)^{-1}(T - S)f_T$. Replace f_T on the right-hand side by $f_T = \tfrac{1}{\lambda}(g + T f_T)$:

$$f_T - f_S = \tfrac{1}{\lambda}(\lambda I - S)^{-1}(T - S)(g + T f_T)$$
$$= \tfrac{1}{\lambda}(\lambda I - S)^{-1}(T - S)(g + T f_S) + \tfrac{1}{\lambda}(\lambda I - S)^{-1}(T - S)T(f_T - f_S).$$

This proves

$$f_T - f_S = \left[\lambda I - (\lambda I - S)^{-1}(T - S)T \right]^{-1} \tfrac{1}{\lambda}(\lambda I - S)^{-1}(T - S)(g + T f_S).$$

Norm estimates yield the assertion. □

The key assumption of Theorem 8.25 is that $\| (T - S)T \|$ is small enough. Set $S = K$ and replace the fixed operator T by a sequence $\{K_n\}$. Then we have to take care that $(K - K_n)K_n \to 0$. This will be achieved by the following definition of collectively compact operators (cf. Anselone [1]).

Definition 8.26 (collective compactness). *A sequence* $\{K_n\} \subset \mathcal{L}(X, X)$ *is called collectively compact if the following set is precompact:*

$$\{K_n \varphi : \varphi \in X, \|\varphi\| \leq 1, n \in \mathbb{N}\}. \tag{8.13}$$

Since we have introduced collective compactness to analyse the Nyström method, we have to demonstrate that the Nyström method produces collectively compact operators K_n. The following assumption that k is continuous is not restrictive, since a quadrature method requires point evaluations.

Lemma 8.27. *Assume that D is compact and $k \in C(D \times D)$. Then any Nyström method based on a stable quadrature rule yields a collectively compact set $\{K_n\}$ of operators.*

Proof. (i) Stability of the quadrature rule in (8.7) is expressed by the inequality $\sum_{k=1}^{n} |a_{k,n}| \leq C_{\text{stab}}$.

(ii) The set (8.13) is uniformly bounded: $|\sum_{k=1}^{n} a_{k,n} k(\cdot, \xi_{k,n}) \varphi(\xi_{k,n})| \leq \sum_{k=1}^{n} |a_{k,n}| |k(\cdot, \xi_{k,n})| |\varphi(\xi_{k,n})| \underset{\text{Part i}, \|\varphi\| \leq 1}{\leq} C := C_{\text{stab}} \|k\|_\infty$.

(iii) Since $D \times D$ is compact, $k(\cdot, \cdot)$ is uniformly continuous. In particular, for all $\varepsilon > 0$ there is some $\delta_\varepsilon > 0$ such that $x, y \in D$ with $\|x - y\| \leq \delta_\varepsilon$ satisfy $|k(x, \xi_{k,n}) - k(y, \xi_{k,n})| \leq \varepsilon / C_{\text{stab}}$. Hence the elements of (8.13) are equicontinuous:

$$\left| \sum_{k=1}^{n} a_{k,n} k(x, \xi_{k,n}) \varphi(\xi_{k,n}) - \sum_{k=1}^{n} a_{k,n} k(y, \xi_{k,n}) \varphi(\xi_{k,n}) \right|$$

$$\leq \sum_{k=1}^{n} |a_{k,n}| |k(x, \xi_{k,n}) - k(y, \xi_{k,n})| |\varphi(\xi_{k,n})| \leq C_{\text{stab}} \cdot \varepsilon / C_{\text{stab}} = \varepsilon.$$

By Theorem 3.51 (Arzelà–Ascoli), the set (8.13) is precompact; i.e., $\{K_n\}$ is collectively compact. $\square$

We show that the operator norm convergence, which is valid for the previous discretisations, implies the new property.

Lemma 8.28. *Suppose that each operator K_n is compact (e.g., because of a finite-dimensional range). If $\|K - K_n\| \to 0$, then $\{K_n\}$ is collectively compact.*

Proof. (i) Remark 8.4 states that K is compact.

(ii) We have to show that $M := \{K_n \varphi : \varphi \in X, \|\varphi\| \leq 1, n \in \mathbb{N}\}$ is precompact. A sequence in M has the form $\{K_{n(j)} \varphi_j\}$ with $n(j) \in \mathbb{N}$ and $\|\varphi_j\| \leq 1$. Let $N := \limsup n(j) \in \mathbb{N} \cup \{\infty\}$.

Case A: $N < \infty$. The subsequence can be chosen such that $n_k := n(j_k) = N$ for all $k \in \mathbb{N}$. By the compactness of K_N, the sequence $K_N \varphi_{j_k} = K_{n_k} \varphi_{j_k}$ has a convergent subsequence.

Case B: $N = \infty$. A subsequence can be chosen such that $n_{k+1} > n_k$. By the same argument as before, there is a convergent subsequence $K \varphi_{j_k}$. Since

$\|K - K_{n_k}\| \to 0$ implies $(K - K_{n_k})\varphi_{j_k} \to 0$, the same subsequence $\{j_k\}$ yields a convergent sequence $K_{n_k}\varphi_{j_k} = K\varphi_{j_k} - (K - K_{n_k})\varphi_{j_k}$. □

Next, we list some properties of collectively compact operators.

Lemma 8.29. *Let* $\{K_n\}$ *be collectively compact.*
(a) Each K_n *is compact.*
(b) If $\{K_n\}$ *is consistent with respect to* K *(i.e.,* $K_n\varphi \to K\varphi$ *for all* $\varphi \in X$*), then* K *is compact,* $\{K_n\}$ *is uniformly bounded, and*

$$\|(K - K_n)K\| \to 0 \quad and \quad \|(K - K_n)K_n\| \to 0.$$

Proof. (i) Let M be the set defined in (8.13). Part (a) follows immediately, since $\{K_n\varphi : \varphi \in X, \|\varphi\| \le 1\} \subset M$ for any n.

(ii) As $K\varphi = \lim K_n\varphi$, it belongs to the closure $\overline{M}$ which is compact. Therefore, K is compact.

(iii) $\{K_n\}$ is uniformly bounded because of Corollary 3.39.

(iv) The image of $E := \{\varphi \in X, \|\varphi\| \le 1\}$ under K or K_n is contained in the compact set $\overline{M}$. Hence, Lemma 3.49 and the definition of the operator norm prove that $\|(K - K_n)K\| = \sup_{\varphi \in E} \|(K - K_n)K\varphi\| \le \sup_{\psi \in \overline{M}} \|(K - K_n)\psi\| \to 0$ and similarly for $(K - K_n)K_n$. □

Exercise 8.30. *Prove: (a) If* $B \subset X$ *is bounded and* $\{K_n\}$ *collectively compact, then* $\{K_n f : f \in B, n \in \mathbb{N}\}$ *is precompact.*
(b) If $\{A_n\}$ *is collectively compact and* $\{B_n\}$ *uniformly bounded (a weaker condition is collective compactness; cf. Lemma 8.29), then* $\{A_n B_n\}$ *is collectively compact.*

The statement of Exercise 8.30b corresponds to the fact that AB is compact if A is compact and B bounded. The conditions on A and B can be interchanged: AB is compact if A is bounded and B compact. The analogue for collective compactness is given in the next lemma. Note that consistency of $\{A_n\}$ implies uniform boundedness (cf. Exercise 8.6a).

Lemma 8.31. *Suppose that* $\{A_n\}$ *is consistent, while* $\{B_n\}$ *is collectively compact. Then* $\{A_n B_n\}$ *is collectively compact.*

Proof. We have to show that the set $M := \{A_n B_n f : \|f\| \le 1, n \in \mathbb{N}\}$ is precompact. A sequence from M has the form

$$\varphi_k := A_{n_k} B_{n_k} f_k \quad \text{with } \|\varphi_k\| \le 1 \text{ and } n_k \in \mathbb{N}.$$

Set $g_k := B_{n_k} f_k$. By collective compactness of $\{B_n\}$, there is a subsequence (denoted again by k) such that

$$g_k = B_{n_k}\varphi_k \to g.$$

Next, two cases must be distinguished: (A) $\sup n_k < \infty$ and (B) $\sup n_k = \infty$.

Case A: We can select a subsequence that such $n_k = n$ is constant; i.e., $g_k = B_n \varphi_k \to g$. Continuity of A_n proves $\varphi_k = A_n g_k \to A_n g$; i.e., a convergent subsequence is found.

Case B: Then $\varphi_k = A_{n_k} g_k$, where, without loss of generality, we may assume $n_{k+1} > n_k$. By consistency, there is some A with

$$A_{n_k} g \to Ag \quad \text{and} \quad \sup_{n_k \in \mathbb{N}} \|A_{n_k}\| < \infty.$$

Therefore, $A_{n_k} g_k - Ag = A_{n_k} (g_k - g) + (A_{n_k} - A)g \to 0$ shows that $A_{n_k} g_k$ is a convergent subsequence. □

Combining Theorem 8.25 and Lemma 8.29, we obtain the following main result.

Theorem 8.32. *Let $\lambda \neq 0$ satisfy[5] $(\lambda I - K)^{-1} \in \mathcal{L}(X, X)$. Suppose that $\{K_n\}$ is consistent with respect to K and collectively compact. Then $\{K_n\}$ is stable and convergent. Furthermore, there is some $n_0 \in \mathbb{N}$ such that for all $n \geq n_0$ the following statements hold:*

$$\|(\lambda I - K_n)^{-1}\| \leq \frac{1 + \|(\lambda I - K)^{-1}\| \|K_n\|}{|\lambda| - \|(\lambda I - K)^{-1}\| \|(K - K_n)K_n\|} \qquad (n \geq n_0);$$

the solutions $f = (\lambda I - K)^{-1} g$ and $f_n = (\lambda I - K_n)^{-1} g$ satisfy

$$\|f - f_n\| \leq \|(\lambda I - K)^{-1}\| \frac{\|(K - K_n)K_n\| \|f\| + \|K - K_n\| \|g\|}{|\lambda| - \|(\lambda I - K)^{-1}\| \|(K - K_n)K_n\|},$$

$$\|f - f_n\| \leq \|(\lambda I - K_n)^{-1}\| \|(K - K_n)f\| \qquad (n \geq n_0).$$

Conclusion 8.33. *Assume that D is compact and $k \in C(D \times D)$. Let $\lambda \neq 0$ be a regular value of K. Then, any Nyström method based on a consistent and stable quadrature is also stable and convergent.*

Proof. By Lemma 8.27 $\{K_n\}$ is collectively compact. Consistency of the quadrature method implies consistency of $\{K_n\}$. Now, Theorem 8.32 yields the statement. □

8.5 Perturbation Results

Lemma 8.34. *Let $\{K_n\}$ be stable with constant C_{stab} : $\|(\lambda I - K_n)^{-1}\| \leq C_{\text{stab}}$. Let $\{T_n\}$ be bounded by $\|T_n\| \leq \zeta/C_{\text{stab}}$ with $\zeta < 1$ for all $n \geq n_0$. Then, also $\{K_n + T_n\}$ is stable.*

Proof. The proof follows the same lines as in Remark 8.8. □

---

[5] In Lemma 8.40 we shall show that this assumption is necessary.

Theorem 8.35. *Suppose that $\lambda \neq 0$ is a regular value of $K + T$; i.e., the inverse $(\lambda I - K - T)^{-1} \in \mathcal{L}(X, X)$ exists. Assume that $\{K_n\}$ is convergent and consistent with respect to $K \in \mathcal{L}(X, X)$, $\{T_n\}$ is collectively compact with $T_n \varphi \to T\varphi$ for all $\varphi \in X$. Then, also $\{K_n + T_n\}$ is stable, consistent, and convergent.*

Proof. (i) By Theorem 8.16b, $\lambda I - K$ is bijective and $(\lambda I - K)^{-1} \in \mathcal{L}(X, X)$ exists.

(ii) To prove that $\{(\lambda I - K_n)^{-1} T_n\}$ is collectively compact, we apply Lemma 8.31 with $A_n = (\lambda I - K_n)^{-1}$ and $B_n = T_n$. By Theorems 8.16 and 8.17a, A_n is consistent with respect to $A = (\lambda I - K)^{-1}$; i.e., the suppositions of Lemma 8.31 are satisfied.

(iii) For any $\varphi \in X$, Theorem 8.17b with $g_n := T_n \varphi \to T\varphi$ proves that $C_n := (\lambda I - K_n)^{-1} T_n$ satisfies $C_n \varphi \to A\varphi$, where $C := (\lambda I - K)^{-1} T$. This shows that $\{A_n\}$ is consistent with respect to A.

(iv) Because of $\lambda I - K - T = (\lambda I - K)(I - C)$ and (i), also the inverse $(I - C)^{-1} \in \mathcal{L}(X, X)$ exists. Since $\{C_n\}$ is collectively compact (cf. (ii)), and consistent with respect to C (cf. (iii)) with $(I - C)^{-1} \in \mathcal{L}(X, X)$, Theorem 8.32 states stability: $\left\| (I - C_n)^{-1} \right\| \leq C$. From

$$(\lambda I - K_n - T_n)^{-1} = (I - C_n)^{-1}(\lambda I - K_n)^{-1}$$

we conclude that also $\{K_n + T_n\}$ is stable.

(v) Consistency of $\{K_n + T_n\}$ is trivial. Since $\lambda I - K - T = (\lambda I - K)(I - C)$ is bijective (cf. (iv)), convergence follows from Theorem 8.17a. $\square$

8.6 Application to Eigenvalue Problems

So far, we have required that $(\lambda I - K)^{-1}$ and $(\lambda I - K_n)^{-1}$, at least for $n \geq n_0$, exist. Then, $\lambda \in \mathbb{C}$ is called a *regular value* of K and K_n. The singular values of K $[K_n]$ are those for which $\lambda I - K$ $[\lambda I - K_n]$ is not bijective. The Riesz–Schauder theory states that for compact K, all singular values are either $\lambda = 0$ or eigenvalues, which means that there is an eigenfunction f such that

$$\lambda f = Kf \qquad (0 \neq f \in X). \tag{8.14a}$$

Similarly, we have discrete eigenvalue problems

$$\lambda_n f_n = K_n f_n \qquad (0 \neq f_n \in X, \ n \in \mathbb{N}). \tag{8.14b}$$

The sets of all singular values are the *spectra* $\sigma = \sigma(K)$ and $\sigma_n = \sigma(K_n)$, respectively. The Riesz–Schauder theory states the following.

Remark 8.36. *Either the range of a compact operator is finite-dimensional (implying that σ is a finite set), or $\lambda = 0 \in \sigma$ is the only accumulation point of the spectrum.*

The obvious question is how $\lambda \in \sigma$ and $\lambda_n \in \sigma_n$ are related. The following results require only consistency of $\{K_n\}$ to K and collective compactness of $\{K_n\}$, which holds for all discretisations discussed before.

Theorem 8.37. *Suppose that $\{K_n\}$ is consistent with respect to K and collectively compact. Let $\{\lambda_n\}$ be a sequence of eigenvalues of (8.14b) with corresponding eigenfunction $f_n \in X$ normalised by $\|f_n\| = 1$. Then, there exists a subsequence $\{n_k\}$ such that either $\lambda_{n_k} \to 0$ or $\lambda_{n_k} \to \lambda \in \sigma$. In the latter case, the subsequence can be chosen such that $f_{n_k} \to f$ with f being an eigenfunction of (8.14a).*

Proof. (i) First we show uniform boundedness: $\sup\{|\lambda_n| : \lambda_n \in \sigma_n, n \in \mathbb{N}\} < \infty$. $|\lambda_n| \le \|K_n\|$ follows from $|\lambda_n| = \|\lambda_n f_n\| = \|K_n f_n\| \le \|K_n\| \|f_n\| = \|K_n\|$ because of $\|f_n\| = 1$. Consistency of $K_n \varphi \to K\varphi$ for all $\varphi \in X$ implies that K_n is uniformly bounded: $C := \sup_{n \in \mathbb{N}} \|K_n\| < \infty$. Together, $|\lambda_n| \le C$ follows.

(ii) Since $\{z \in \mathbb{C} : |z| \le C\}$ is compact, there is a convergent subsequence $\lambda_n \to \lambda$ $(n \in \mathbb{N}')$. If $\lambda \ne 0$, we have to show that $\lambda \in \sigma$. Because of $\|f_n\| = 1$, collective compactness of $\{K_n\}$ implies that there is a second subsequence $n \in \mathbb{N}'' \subset \mathbb{N}'$ for which $K_n f_n$ is convergent. By (8.14b) and $\lambda_n \to \lambda \ne 0$, this yields convergence of f_n to some f:

$$f_n = \frac{1}{\lambda_n} K_n f_n \to f.$$

Continuity of the norm yields $\|f\| = 1$, implying $f \ne 0$. The first two terms in $K_n f_n = K_n (f_n - f) + (K_n - K) f + Kf$ vanish (use $\sup_{n \in \mathbb{N}} \|K_n\| < \infty$ and consistency), so that $K_n f_n \to Kf$ as well as $K_n f_n = \lambda_n f_n \to \lambda f$, proving that $\lambda \in \sigma$ is an eigenvalue of K. $\square$

The statement of Theorem 8.37 about the eigenvalues can be abbreviated by 'the set of accumulation points of $\bigcup_{n \in \mathbb{N}} \sigma_n$ is contained in σ'. Now we prove the reverse direction: 'σ is contained in the set of accumulation points of $\bigcup_{n \in \mathbb{N}} \sigma_n$'.

Theorem 8.38. *Suppose that $\{K_n\}$ is consistent with respect to K and collectively compact. Then, for any $0 \ne \lambda \in \sigma$ there is a sequence $\{\lambda_n\}$ of eigenvalues from (8.14b) with $\lambda = \lim_{n \in \mathbb{N}} \lambda_n$. Again, a subsequence can be chosen so that the eigenfunctions f_{n_k} from (8.14b) converge to an eigenfunction of (8.14a).*

We prepare the proof by two lemmata. The following functions on $\mathbb{C}$ will be needed:

$$\varkappa_n(\lambda) := \begin{cases} 1/\|(\lambda I - K_n)^{-1}\| & \text{if } \lambda \notin \sigma_n, \\ 0 & \text{otherwise.} \end{cases}$$

Exercise 8.39. *Prove that $\varkappa_n$ is not only continuous, but also satisfies the Lipschitz property $|\varkappa_n(\lambda) - \varkappa_n(\mu)| \le |\lambda - \mu|$ for all $\lambda, \mu \in \mathbb{C}$.*

The next lemma describes the equivalence of $\lim_{n \in \mathbb{N}} \varkappa_n(\lambda) = 0$ and $\lambda \in \sigma$.

Lemma 8.40. *Suppose that $\{K_n\}$ is consistent with respect to K.*
(a) $\lim_{n\in\mathbb{N}} \varkappa_n(\lambda) = 0$ *for each $\lambda \in \sigma$.*
(b) Suppose, in addition, that the set $\{K_n\}$ is collectively compact, $\lambda \neq 0$, and $\lim_{n\in\mathbb{N}} \varkappa_n(\lambda) = 0$. *Then $\lambda \in \sigma$ is an eigenvalue of K.*

Proof. (i) Let $\lambda \in \sigma$ with corresponding eigenfunction $f \neq 0$: $\lambda f = Kf$. Set $g_n := \lambda f - K_n f$. Consistency implies that $g_n \to 0$. For an indirect proof assume that $\lim_{n\in\mathbb{N}} \varkappa_n(\lambda) \geq 2\varepsilon > 0$ and choose a subsequence such that $\varkappa_n(\lambda) \geq \varepsilon$; i.e., $\|(\lambda I - K_n)^{-1}\| \leq 1/\varepsilon$. Then

$$\|f\| = \|(\lambda I - K_n)^{-1} g_n\| \leq \|(\lambda I - K_n)^{-1}\| \|g_n\| \leq \|g_n\|/\varepsilon \to 0$$

is a contradiction to $f \neq 0$. This ends the proof of Part (a).

(ii) The assumptions of Part (b) imply $\|(\lambda I - K_n)^{-1}\| \to \infty$ (we write formally $\|(\lambda I - K_n)^{-1}\| = \infty$ for $\lambda \in \sigma_n$). Hence, there are functions $\{f_n\}$ and $\{g_n\}$ with

$$(\lambda I - K_n) f_n = g_n, \quad \|f_n\| = 1, \quad g_n \to 0.$$

By collective compactness, a subsequence $\{K_{n_k} f_{n_k}\}$ is convergent. Therefore $f_{n_k} = (g_{n_k} + K_{n_k} f_{n_k})/\lambda$ has some limit f with

$$\|f\| = 1 \quad \text{and} \quad f = \lim(g_{n_k} + K_{n_k} f_{n_k})/\lambda = Kf/\lambda;$$

i.e., λ is an eigenvalue of K (details as in Part (ii) of the proof above). $\square$

Lemma 8.41. *Let $\Omega \subset \mathbb{C}\backslash\{0\}$ be compact. Then, either $\sigma_n \cap \Omega \neq \emptyset$ or $\varkappa_n$ takes its minimum on the boundary $\partial\Omega$.*

Proof. Assume that $\sigma_n \cap \Omega = \emptyset$ and fix some $\varphi \in X$ with $\|\varphi\| = 1$. By assumption, $(zI - K_n)^{-1}\varphi$ is well-defined for all $z \in \Omega$. Since $\varkappa_n$ is continuous, and Ω is compact, $\max_{z\in\Omega} \varkappa_n(z) = \varkappa_n(\lambda)$ holds for some $\lambda \in \Omega$. Choose $\Phi \in X^*$ according to Corollary 3.11 such that $\|\Phi\|_X^* = 1$ and $\Phi((\lambda I - K_n)^{-1}\varphi) = \|(\lambda I - K_n)^{-1}\varphi\|$, and define a complex function by

$$F(z) := \Phi((zI - K_n)^{-1}\varphi) \in \mathbb{C}.$$

Its derivative $F'(z) = -\Phi((zI - K_n)^{-2}\varphi)$ is defined in Ω, so that F is holomorphic in Ω. Hence, $|F|$ takes its maximum on $\partial\Omega$:

$$\|(\lambda I - K_n)^{-1}\varphi\| = F(\lambda) = |F(\lambda)| \leq \max\{|F(z)| : z \in \partial\Omega\}.$$

The inequality

$$|F(z)| = |\Phi((zI - K_n)^{-1}\varphi)| \leq \|\Phi\|_X^* \|(zI - K_n)^{-1}\varphi\|$$
$$\leq \|(zI - K_n)^{-1}\| = 1/\varkappa_n(z)$$

holds for all $z \in \partial\Omega$, so that

$$\|(\lambda I - K_n)^{-1}\varphi\| \leq \max\{1/\varkappa_n(z) : z \in \partial\Omega\} = 1/\min\{\varkappa_n(z) : z \in \partial\Omega\}.$$

Since the choice of $\varphi \in X$ with $\|\varphi\| = 1$ is arbitrary,

$$\|(\lambda I - K_n)^{-1}\| \leq \frac{1}{\min\{\varkappa_n(z) : z \in \partial\Omega\}}$$

follows, which is equivalent to $\min\{\varkappa_n(z) : z \in \partial\Omega\} \leq \varkappa_n(\lambda)$; i.e., the minimum is taken on $\partial\Omega$. $\quad\square$

Now, we give the proof of Theorem 8.38.

(i) K is compact (cf. Remark 8.29b), and each eigenvalue $0 \neq \lambda \in \sigma$ is isolated (Remark 8.36). Choose a complex neighbourhood Ω of λ such that

$$0 \notin \Omega \subset \mathbb{C}, \quad \Omega \cap \sigma = \{\lambda\}, \quad \lambda \in \Omega\backslash\partial\Omega, \quad \Omega \text{ compact.} \tag{8.15}$$

For $\alpha_n := \min\{\varkappa_n(z) : z \in \partial\Omega\}$ we want to show $\alpha_n \geq \varepsilon > 0$ for $n \geq n_0$. For an indirect proof, assume $\alpha_{n_k} \to 0$. Define $z_n \in \partial\Omega$ by $\alpha_n = \varkappa_n(z_n)$. Since $\partial\Omega$ is compact, $z_{n_k} \to \zeta$ for some $\zeta \in \partial\Omega$. By $|\varkappa_{n_k}(\zeta) - \varkappa_{n_k}(z_{n_k})| \leq |\zeta - z_{n_k}|$ (cf. Exercise 8.39), $\varkappa_{n_k}(z_{n_k}) \to 0$ follows and implies that $\zeta \in \Omega \cap \sigma$ (cf. Lemma 8.40b) in contradiction to (8.15).

(ii) Assume $\alpha_n \geq \varepsilon > 0$ for $n \geq n_0$. As $\varkappa_n(\lambda) \to 0$ (cf. Lemma 8.40a), one can choose n_0 such that $\varkappa_n(\lambda) \leq \varepsilon/2$ for $n \geq n_0$. Since the minimum is not attained on $\partial\Omega$, one concludes from Lemma 8.41 that $\Omega \cap \sigma_n \neq \emptyset$; i.e., there are eigenvalues $\lambda_n \in \sigma_n$ for all $n \geq n_0$. Since Ω can be an arbitrarily small neighbourhood, there are $\lambda_n \in \sigma_n$ with $\lim \lambda_n = \lambda$. Theorem 8.37 yields the statement about the eigenfunctions.

References

1. Anselone, P.M.: Collectively Compact Operator Approximation Theory and Applications to Integral Equations. Prentice-Hall, Englewood Cliffs (1971)
2. Brakhage, H.: Über die numerische Behandlung von Integralgleichungen nach der Quadraturformelmethode. Numer. Math. **2**, 183–196 (1960)
3. Fredholm, E.I.: Sur une classe d'équations fonctionnelles. Acta Math. **27**, 365–390 (1903)
4. Hackbusch, W.: Tensor spaces and numerical tensor calculus, *Springer Series in Computational Mathematics*, Vol. 42. Springer, Berlin (2012)
5. Hsiao, G.C., Wendland, W.L.: Boundary Integral Equations. Springer, Berlin (2008)
6. Linz, P.: A general theory for the approximate solution of operator equations of the second kind. SIAM J. Numer. Anal. **14**, 543–554 (1977)
7. Nyström, E.J.: Über die praktische Auflösung von linearen Integralgleichungen mit Anwendungen auf Randwertaufgaben der Potentialtheorie. Soc. Sci. Fenn. Comment. Phys.-Math. **4**(15) (1928)
8. Sauter, S.A., Schwab, C.: Boundary element methods, *Springer Series in Computational Mathematics*, Vol. 39. Springer, Berlin (2011)
9. Schmidt, E.: Zur Theorie der linearen und nichtlinearen Integralgleichungen. I. Teil: Entwicklung willkürlicher Funktionen nach Systemen vorgeschriebener. Math. Ann. **63**, 433–476 (1907)
10. Yosida, K.: Functional Analysis. Springer, Berlin (1968)

Index

W. Hackbusch, *The Concept of Stability in Numerical Mathematics*,
Springer Series in Computational Mathematics 45, DOI 10.1007/978-3-642-39386-0,
© Springer-Verlag Berlin Heidelberg 2014

Printed in the United States
By Bookmasters